Logic, Argumentation & Reasoning

Interdisciplinary Perspectives from the Humanities and Social Sciences

Volume 20

Logic, Argumentation & Reasoning (LAR) explores links between the Humanities and Social Sciences, with theories (including decision and action theory) drawn from the cognitive sciences, economics, sociology, law, logic, and the philosophy of science.

Its main ambitions are to develop a theoretical framework that will encourage and enable interaction between disciplines, and to integrate the Humanities and Social Sciences around their main contributions to public life, using informed debate, lucid decision-making, and action based on reflection.

- Argumentation models and studies
- Communication, language and techniques of argumentation
- Reception of arguments, persuasion and the impact of power
- Diachronic transformations of argumentative practices

LAR is developed in partnership with the Maison Européenne des Sciences de l'Homme et de la Société (MESHS) at Nord - Pas de Calais and the UMR-STL: 8163 (CNRS).

Proposals should include:

- A short synopsis of the work, or the introduction chapter
- The proposed Table of Contents
- The CV of the lead author(s)
- If available: one sample chapter

We aim to make a first decision within 1 month of submission. In case of a positive first decision, the work will be provisionally contracted—the final decision about publication will depend upon the result of an anonymous peer review of the complete manuscript.

The complete work is usually peer-reviewed within 3 months of submission.

LAR discourages the submission of manuscripts containing reprints of previously published material, and/or manuscripts that are less than 150 pages / 85,000 words.

For inquiries and proposal submissions, authors may contact the editor-in-chief, Shahid Rahman at: shahid.rahman@univ-lille3.fr, or the managing editor, Juan Redmond, at: juan.redmond@uv.cl

More information about this series at http://www.springer.com/series/11547

Nicholas Rescher

Luck Theory

A Philosophical Introduction to the Mathematics of Luck

Nicholas Rescher
University of Pittsburgh
Pittsburgh, PA, USA

ISSN 2214-9120 ISSN 2214-9139 (electronic)
Logic, Argumentation & Reasoning
ISBN 978-3-030-63782-8 ISBN 978-3-030-63780-4 (eBook)
https://doi.org/10.1007/978-3-030-63780-4

This Springer imprint is published by the registered company Springer Nature Switzerland AG
The registered company address is: Gewerbestrasse 11, 6330 Cham, Switzerland

For Bas van Fraassen
Philosopher par excellence

Preface

One may well ask: Where in the taxonomy of branches of knowledge does the theory of luck belong? Is it to philosophy—and if so, to metaphysics or epistemology or ethics—or is it to some branch of applied mathematics such as economics or decision theory, or is it a matter of psychology and personality study? The answer is simple: it belongs to none but is entangled with all. It is as interdisciplinary a branch of investigation and inquiry as there is or can be. Clearly no single book can do justice to the subject. All that is attempted here is to lay the groundwork for the mensurational and mathematical approach to the comprehension of luck which affords otherwise unavailable. The aim is to inject clarity and precision into our understanding of these issues. But unless we can see time and again that calculation harmonizes with intuition, the validating of our proceedings remains in doubt. Fortunately, however, we shall find that all is well in this regard.

Overall, the academic study of luck has three main branches:

- *The philosophy of luck.* This project was addressed in my book, *Luck: The Brilliant Randomness of Everyday Life* (New York: Farrar-Straus-Giroux, 1995; reprinted in 2000 by the University of Pittsburgh Press). Further discussion and references to the vast literature can be found in Ian M. Church and Robert J. Hartman (eds.), *The Routledge Handbook of the Philosophy of Luck* (New York and London: Routledge, 2019).
- *The psychology of luck.* This project is addressed in N. N. Taleb, *Fooled by Randomness: The Hidden Role of Chance in Life* (New York: Random House, 2004). (See also the preceding entry.)
- *The mathematics of luck.* This heretofore undealt-with project of developing a metric approach to luck is the topic of this book.

For further details regarding the extensive literature on luck, see the Bibliography at the end of this book.

The English term "luck" covers a good deal of territory. There is *stochastic* luck of matters governed by chance issue in success or failure: playing dice, hitting the jackpot, or striking oil. This is at issue when we say small things as: "He had good luck at the races" or "He thanked his lucky stars for having succeeded." However, there is also the *existential* (or *constitutive*) luck of being treated well by fate and fortune. This is at issue when we say such things as: "She has the good luck of being born to wealthy parents" or "He had the bad luck of coming of age just in time to be caught up in the wartime draft." It is the former, chance-outcome version of luck that will be at the forefront of the present deliberations.

In analyzing chancy situations, luck and risk play different yet complementary roles. Luck dominates in the region outside the range of expectation. Risks, by contrast, prevail in the negative sector of this range. And both stand in contrast to the hopefully anticipant positivity.

As the Bibliography at the book's end shows, recent years have seen numerous informative studies of the psychology of luck beliefs and believers.[1] The result findings indicate such conclusions as that the young tend to feel luckier than the old, and that young female students tend to feel luckier than young males. However, such psychological and sociological considerations are irrelevant to the present range of concerns, which are addressed not to the extent to which people *feel* lucky, but rather to the assessment of the extent to which they can objectively be deemed actually to be so.

In its stochastic aspect, luck is a matter of the yields and likelihoods of the possible outcomes of chancy situations. Accordingly, both value theory and probability theory are inextricably involved in the theory of luck. Both of these pivotal factors—both yields and probabilities—can be assessed either via personal and subjective judgments or via the objective and impersonal proceedings at issue in rational decision theory.

Luck theory aims at a formularization of the conception whose appropriateness is attested by the fact that virtually all consequences of the theory are intuitively acceptable and that, conversely, virtually all intuitively acceptable features of luck can be verified in the theory.[2] It aims at an adequation that harmonizes formalism and informal understanding. In the end, the theory should make good a claim to providing for a mathematically precise systematization of an informal and pre-systematic conception.

I am grateful to Estelle Burris for her conscientious help in preparing this book for the press.

Notes

1. Davies 1997 is a typical instance of a paper on belief in luck. (References of this format relate to the Bibliography.)
2. In the present context, "virtually all" comes to "all save some few explicable exceptions."

Pittsburgh, PA, USA
August 2020

Nicholas Rescher

Philosophical Prologue (Fortune, Fate, and Chance)

We live in a world where our aims and goals, our "best laid plans," and indeed our very lives are at the mercy of fortuitous chance and inscrutable contingency. In such a world, where we propose and fate disposes, where the outcomes of all too many of our actions depend on "circumstances beyond our control," luck is destined to play a leading role in the human drama.[3]

The role of chance in human affairs was once the topic of extensive discussion and intensive debate among philosophers. In Hellenistic Greece, theorists debated tirelessly about the role of *eimarmenê*, the unfathomable fate that remorselessly ruled the affairs of the men and gods alike, regardless of their wishes and actions. The Church fathers struggled mightily to combat the siren appeal of the ideas of chance and destiny—those superstition-inviting potencies. (Saint Augustine detested the very word *fate*.) The issue of good or bad fortune, along with the related question of the extent to which we can control our destinies in this world, came to prominence again in the Renaissance, when scholars brooded once again about the issues of human destiny raised by Cicero and Augustine. Calvinism kept many a theologian and philosophers awake at night. And the topic undoubtedly has a long and lively future before it, since it is certain that, as long as human life continues, luck will play a prominent part in its affairs.

Luck as an English word is a creature of the fifteenth century and derives from the Middle High German *gelücke* (modern German *Glück*), which (somewhat unfortunately) means both happiness and good fortune, conditions which, given human fecklessness, are certainly not necessarily identical. Virtually from its origin, the term has been applied particularly to good or ill results in gambling, games of skill, and chancy ventures in general![4]

It is useful for the range of present deliberations to have (what several European languages do not make available!) a single word to mean "good or bad fortune acquired unwittingly, by accident or chance." In English, "luck" does this job quite well; in other languages, we have to do the best we can.[5] For luck fares rather mixedly in European languages. The Greek *tuchê* is too much on the side of haphazard. In Latin, *fortuna* comes close to its meaning, with the right mixture of chance (*casus*) and benefit. But the German suffers from the unfortunate equivocation that *Glück* means not only luck (*fortuna*) but also happiness (*Felicitas*). The French "chance" (from the Latin *cadere*, meaning how matters fall out—"how the dice fall") is a fairly close equivalent of luck, however. And the Spanish *suerte* is also pretty much on target.

On the other side of the coin, several languages have a convenient one-word expression for "a piece of bad luck" (French *malchance*, German *Pech*)—a most useful resource considering the nature of things, which English unaccountably lacks. (Despite its promising etymology, misfortune is not quite the same, since it embraces any sort of mishap, not merely those due to unpredictable accident or chance but also those due to one's own folly or to the malignity of others.) And it may be emblematic of something larger that no European language seems to have a one-word expression for "a piece of good luck."

Luck is not a causal force or function in nature; indeed it is not a natural phenomenon at all. Rather it is a humanly contained conception imposed by us on occurrences to make the course of events easier to describe and understand. (In this regard, it is like degrees of latitude or days of the week.)

* * *

Luck requires a potential beneficiary/maleficiary—someone who has an interest in the matter. So understood, someone is lucky (or unlucky) when they are the beneficiary (or maleficiary) of a fortuitous development. As such, good or bad luck can take very different forms, some standard versions being:

- *Finder's luck*: stumbling upon a treasure trove
- *Gambler's luck*: succeeding or failing in some sort of gamble
- *Guesser's luck*: inadvertently hitting on the right answer (e.g., in a spelling bee)
- *"Dumb" luck* (good or bad): being in the right (or wrong) place at the right (or wrong) time: e.g., stumbling out of the way just as the bullet comes by, and similar "narrow escapes"

Each such mode of luck has characteristic features of its own, so that none is totally typical of the entire range. However, the present deliberations will focus on gambler's luck, both because it is the most familiar form of the phenomenon and because it best admits of quantitative analysis. Luck requires chance. When a development eliminates all your superior risk from a competition you are lucky that this has occurred. But your examining is now no longer something that occurs by luck. A chancy outcome that is lucky may not be beneficial or particularly welcome when it qualifies only because all alternatives are worse. And even the best-possible outcome will not necessarily be lucky—specifically when all the alternatives have the same yield.

* * *

It is instructive to consider some of the ways and means of luck in various areas of human endeavor. For luck takes many forms and is able to adjust its coloration, chameleon-like, to the background of various settings.

(1) *Luck in Games*. Success or failure in situations of competition and conflict often hinges on matters of fortuitous happenstance. Even in conflicts of skill rather than chance—especially in sporting and gaming competitions—luck plays an enormous role. A player's momentary distraction here or accidental slip-up there can open up the opportunity to add a decisive point to the score. When the opposing team's star player happens "to have an off day," our team may win the point that gains the championship. It is precisely because interesting games are those between evenly matched teams, where victory is not a foregone conclusion owing to an imbalance of skill, that the role of chance eventuations—and thus of luck—is so prominent in professional sport. And it is an easy step from sport to the more serious conflicts at issue with luck in warfare.

(2) *Luck in Warfare*. There is enormous scope for luck in warfare. An accidentally intercepted message may betray plans and intentions to the enemy, and a tactical maneuver gone wrong through a fluke may create a decisive opportunity for the opponent in battle. The fog of unknowing that covers the battlefield opens doors beyond number for the entry of luck. And in war, timing is everything. It was a piece of very bad luck indeed for Robert E. Lee at Gettysburg that Gen. J. E. B. Stuart decided to take his cavalry off raiding instead of providing scouting cover for the invading Confederate army. It was a piece of very good luck indeed for the Americans at Yorktown that the British under Cornwallis made their foray before the French fleet under De Grasse had to return to its winter station.

A mode of warfare somewhat different from the military is the political. Here, too, there is ample scope for luck, as, for example, with luck in elections.

(3) *Luck in Elections*. The democratic electoral process obviously affords enormous scope for the operation of luck. A candidate's ill-timed flu can re-open large questions about his state of health at an awkward time. Bad weather on election day may favor a liberal candidate by keeping the more conservative elderly voters at home. In the 1890–1930 era, Democratic candidates did not have much of a chance in US presidential elections, but Teddy Roosevelt's Bull Moose split of the Republican party in 1912 created a crack wide enough to let Woodrow Wilson slip through. One of the cardinal reasons why "A week is a long time in politics" (as the saying goes) is that fortuitous developments here have the prospect of engendering a lot of elbow room for luck.

(4) *Luck in Search and Research*. Be it in prospecting or in scientific investigations, many search processes have a hit-and-miss aspect that allows substantial room for the operation of luck. Scientific discoveries are often made not on the basis of some well-contrived plan of investigation but through some stroke of sheer luck—a phenomenon common enough that a specific name has sprung up for it, such discoveries being said to be made "by serendipity."[6] This occurs in science when investigators come upon answers to questions or solutions to

problems by haphazard chance rather than by design, planning, contrivance, and the use of methods. Consider, for example, such dramatic instances as Henri Becquerel and his photo plates in the context of the discovery of radioactivity, and Alexander Fleming and his yeast mold in the context of the discovery of antibiotics. The numberless instances of this general phenomenon show that luck is not only a prominently operative factor in such explicitly chancy matters as gambling or risk-taking entrepreneurship, but that it also plays a significant role in such thoroughly rational enterprises as scientific inquiry. And there is no reason for disparagement here. After all, serendipitous discoveries are still discoveries.

(5) *Luck in Information and Knowledge*. Epistemic luck is not confined to science but is a general phenomenon. Convinced that Mary is next door, you claim: "I know that there is a tall woman next door." But unbeknownst to you, Mary has left the room, and Jane, who is also tall, has entered it. Is your claim correct? Your belief that "There is a tall woman next door" is true all right—no question about it. But the fact at issue—that there is a tall woman next door—and the ground for your belief in it (namely, that it is Mary who is next door) have become totally disconnected. It is only *by accident* that your belief is true. To be sure, this is not good enough for actual *knowledge*, as this term is generally understood. To constitute genuine knowledge, beliefs must not just be true—they must be properly grounded as well.

(6) *Moral Luck.* When someone answers a question correctly, this will—if resulting by pure guesswork—be a matter of what has come to be called "epistemic luck." For knowledge, it is not enough to have the correct answer by mere "epistemic luck."[7] If someone asks you for the (positive) square root of 81 and you respond 9 under the bizarre impression that square roots are compiled by summing up the digits involved (with $8 + 1 = 9$), the correctness of your response does not mean that you know the answer. Your correct answer did not reflect knowledge. Correct conjecture is a matter of luck. After all, if you do the ethically right thing by chance—e.g., restore something to its proper owner by giving it to a random recipient who "just happens" to be the owner, you earn no credit for having kept a promise. If you *accidentally* pick the right medicine for someone, you get no credit for providing aid. Strictly speaking, there is no such thing as "moral luck."

The variation of luck makes this a diversified cluster of local considerations.

I can make a bet with *A* and bet with *B* that I will lose the former. Winning the second means losing the first, but can turn bad luck fortune into good. Reasoning about luck can become convoluted.

Someone may be unlucky in getting himself into a fix and lucky to get out of it cheaply. Thus, consider: Within the circumstances, *X* could have crossed the Atlantic in any of four sailings but chose the Titanic. Only 20 percent of the first class males survived. *X*'s choice was unlucky, his fate very lucky. For since we want to manage our affairs in such a way that we can benefit from good luck and can avert the losses that result from bad luck. John is a middling tennis player and generally turns in a mediocre performance. But at golf he occasionally "punches beyond his weight" and while usually mediocre occasionally wins against far better players. His occasional exceeding of expectation means that John is luckier at golf than at tennis. But does this make him luckier overall? Might he not be better off if his records at golf and tennis were interchanged?

* * *

Suppose that someone is intent on murder but finds that just when he is about to pull the trigger his intended victim falls dead with a heart attack. One would certainly not relieve him of *moral* reprehension (whatever the legal status). Or alternatively, if in striving to save a drowning child, our agent found that another, stronger swimmer had unforeseeably arrived a moment sooner. We would then surely not withhold approbation through deficiency in what is called "moral luck," which has been the subject of considerable debate.[8]

* * *

Chance is the Joker in human affairs. Fate's powerful tool. "Pure chance and nothing but chance—absolute, blind, accident" is what provides "the basis of the wondrous structure of evolution" writes Jacques Monod.[9] And if the evolutionary biologists of the present day have it right, the operation of chance is nowhere more prominent than in the biological sphere. As they see it, evolution is driven by an interaction between chance and capacity, between random genetic mutations on the one hand and elimination by natural selection on the other. George G. Simpson has vividly stressed that many chancy twists and turns lie along the evolutionary road, insisting that "the fossil record shows very clearly that there is no central line leading steadily, in a goal-directed way, from a protozoan to man . . . If the causal chain had been different, *Homo sapiens* would not exist."[10] Biological evolution is clearly a chancy business. The movement of the rainclouds may determine the survival of a biological species and thus makes for a crucial evolutionary difference not only for its heirs, but also for its predators and their heirs, its parasites and their heirs, etc. The workings of evolution—be it of life or intelligence or culture or technology or science—are always the product of a great number of individually unlikely events. What is involved is not so much a survival of the fittest as a survival of the fortunate. And insofar as we consider it a good thing that our species is here today in its present form, we have little real choice but to see this as a matter of luck.

The unfolding of evolutionary developments involves putting to nature a series of questions whose successive resolution produces a result reached along a route that traces out one particular contingent path within a space of alternatives that at each step open the way to yet further possibilities. An evolutionary process is thus a very iffy proposition—a complex labyrinth in which a great many twists and turns in the road must be taken right for matters to end up as they do. If the evolutionary game were somehow replayed, even under the selfsame conditions, the outcome produced by the random fluctuations at issue would in all likelihood be completely different.[11]

To characterize evolution as a matter of "the survival of the fittest" is not exactly right—it would not be altogether inappropriate to call it "the survival of the lucky." (Think of the tidal wave that washes most supra-microscopic animals from an island habitat.) Insofar as modern theories of cosmology, biology, and sociology are right, and pure stochastic chance plays the sort of fundamental role in the world's scheme of things that these descriptions envision, it is clear that our existence—be it individual (personal) or collective (species-connected)—is a matter of luck. And so we humans owe to its favor not only the good things and opportunities that life affords us, but even life itself as we know it. Accordingly, insofar as we may categorize our own existence as a benefit—or at least our existence as a biological species—it must be seen as a matter of luck that we exist in this world, be it as individuals or as a biological kind. Our existence as encompassing our past, our present, and our future is unavoidably at the mercy of chance and luck. And without some understanding of the nature of luck and its role, we cannot achieve a proper grasp on the human condition.

* * *

As individuals, we may never know how lucky we actually are. With every step we take, chance can intervene for our good or ill. For all we know, we narrowly escape death a dozen times each day—failing to inhale a fatal microbe here, and there missing by a hair's breadth the pebble that would cause us to slip and pitch into an onrushing bus. Luck, then, is a formidable and ubiquitous factor in human life as we know it—a companion that, like it or not, accompanies us all from the cradle and to the grave.

Luck is at work when things that are of significance to us occur fortuitously, by chance, as it were.[12] And "significance" here means that benefits or negativities must be involved. Sometimes, to be sure, a benefit can be assessed as such only in retrospect. Whether a marriage turns out well or not is something that will not be apparent on the wedding day. Accordingly, whether or not a man and wife are lucky to have found one another will be determinable only with the wisdom of hindsight. Generally, however, we judge goods and evils in the short run, without worrying about "how they will turn out in the end." (After all, as John Maynard Keynes observed, "in the long run we are all dead.")

* * *

Luck pivots on unpredictability. A world in which agents foresee everything to go according to a discernible plan leaves no room for luck. But we ourselves live in a very different sort of world. Things often go well or ill for us due to conditions and circumstances that lie wholly beyond our cognitive or manipulative control. It was a matter of bad luck for the Spain of King Philip II when a storm scattered the "Invincible Armada" in the English Channel. But it was a matter of good luck for Queen Elizabeth's subjects. Luck—good or ill—impinges upon individuals and groups alike (think of the Jews of Poland or the passengers on the *Titanic*). There is no way of escaping it in this world. It is not just that having children is to give hostages to fortune, but having a stake in anything whatsoever. Wherever we invest our hopes and goals and objectives—whatever may be our expectations and aspirations and plans—good or bad luck can come into operation to realize or frustrate our wishes. Our best laid plans, like that of Robert Burns' mouse, "gang aft agley" and do so for reasons entirely beyond our knowledge and control. We play our cards as best we can but the outcome depends on what is done by the other players in the system—be they people or nature's forces. Our lives are lived amidst hopes and apprehensions. Things can turn out for our weal or our woe in ways that we can neither foresee nor control. And it is exactly here that the factor of luck makes its inexorable way into the domain of human affairs. Often as not, a person's life is a chain constructed from links of luck. The youthful personal influences that inform one's career decisions, the contingencies that determine one's employment, the chance encounters that lead to one's marriage, etc. are so many instances of luck.

However, it is folly to think that good luck can be earned or deserved. Good luck may invite envy but it should never elicit admiration since its beneficiary got there by pure chance. Good luck is something to be happy about, but not really something to be grateful for. No larger potency in the scheme of things arranged for it to come your way; chance alone has been at work. So, don't press your luck. Insofar as you can decrease reliance on luck, you can diminish your exposure to risk, but do not count on luck itself to get you out of a tight spot.

* * *

To be sure, a *deterministic* world is one where ontological chance plays no part—would be one that is (in principle) *completely* predictable. But the fact is that recent physics has come to see our world as indeterministic. It emphasizes the role of chance and stochastic randomness in the world's eventuations, stressing those numerous physical phenomena which, like quantum process, are inherently probabilistic. We should not (for example) ask for a prediction of the time when a certain atom of an unstable transuranic element will disintegrate into simpler components, because such a time cannot in principle be predetermined. From the perspective of contemporary physics, the prominence of stochastic processes in nature leads us to turn from exact forecasts to probability distributions. And, of course, the operation of chance is not confined to physics alone. In biology we see randomness at work in genetic mutations, in economics there is the random walk theory of stock market price fluctuation, and so on. Chance is a pervasive factor in modern science, as witness the diffusion of probabilistic and statistical techniques throughout this domain. To be sure, different ranges of phenomena vary sharply in the scope that they provide for chance: it is obviously greater in politics than in celestial mechanics.

To be sure ontological impredictability is not the only sort. For there is also the epistemic impracticability due to information that is incomplete—be it for contingent or for necessary reasons. And of course, impredictability of any manner or description opens a doorway to the entry of luck.

Chance or haphazard is a crucial factor in relation to luck, but it comes in many forms.

- *Natural chance* as inherent in the random or stochastic processes of physical reality and best monitored via relevant laws and regularities of nature (as per the chance of rain, or of a heart attack)
- *Sociological chance* as inherent in the actions of people and best monitored by social statistics (as per the chance of burglary or of murder or suicide)
- *Mechanical chance* inherent in the symmetric functioning of process-symmetric mechanisms and best monitored by design analysis (as per coin tosses, roulette wheels, or card drawings)
- *Epistemic chance* resulting from the incompleteness of knowledge that leaves room for impredictability on the basis of alternatives obtaining "for aught we know"—or even for aught we possibly *can* know— in the circumstances
- *Human haphazard* condition with human fecklessness and impredictability of choice

In consequence there are two importantly different modes of chanciness: that of statistic processes in nature's *modus operandi* (like the "swerve" of Epicurus) and that of informative insufficiency in the cognitive condition of man. And either mode of chanciness suffices to make room for the machinations of luck.

* * *

Luck, the present perspective, obtains for someone a chancy situation when their outcome yield exceeds reasonable expectation. Secure foreseeability eliminates luck. When the handwriting is on the wall so to speak, there is no room for luck about it, for, by hypothesis, the door is now shut to change and choice. To be achieved by luck, an outcome must be realized unforeseeably, fortuitously, and by accident from the conceived agent's point of view: as far as he sees, it is unforeseeable.

But what sort of unpredictability/unforeseeability is at issue?

1. *Logical impredictability*. One cannot (meaningfully) predict today that 2 + 2 = 4 will no longer hold tomorrow.
2. *Physical impredictability*. One cannot predict the inherently random stochastic phenomena of nature such as when a particular transuranic atom will decay. Nor yet can you predict chaotic phenomena such as the evolution of smoke wisps.
3. *Social impredictability*. One cannot predict the indeterminate parameters of individual decision as with the exact number of people in St. Peter's Square at noon tomorrow.
4. *Agency impredictability*. One cannot (securely) predict today the outcome of decisions you will only make tomorrow.
5. *Ignorance impredictability*. One often does not know the relevant facts required for prediction as those ignorant of astronomy cannot predict when an eclipse will occur.[13]

Matters of luck require that—for whatever reason—the outcomes concerned are for aught the agent knows intellectable contingencies whose realization cannot securely be predicted. Agent-relative cognition incapacity is the crux of the chancy impredictability that characterizes luck.

* * *

But a big question lurks in the background. Even if one grants that the preceding analysis adequately characterizes how people *think* of luck in terms of chance, the question yet remains: Is there really any such thing?

There are certainly doctrinal positions that deny the existence of luck. In the main they are of three types:

- *Mechanistic Determinism*: The world is one vast machine of sorts all of whose operations are unavoidingly predetermined by nature's inexorable laws. Determinative atomists such as Lucretius, mechanistic determinists like Laplace, and dialectical materialists like Marx and Engels have held positions of this sort.
- *Metaphysical Determinism*: The world's eventuations are one and all predetermined from the very outset of time by principles of lawful order that necessitates all of its occurrences. The ancient Stoics, Spinoza, and scientistic positivist like Comte have maintained this sort of thing.
- *Theological Predestinationism*: Some have envisioned the world's history as the temporal unfolding of a vast and all-determinative program through which God sets into action an all-determinative plan by which all of history's developments are predetermined with inexorable certainty like the events unfolding when a filmstrip is played. Islamic fatalists and Calvinists are of this persuasion.

None of these positions allows any room for chance, accident, and choice contingency in the world's scheme of things. As they see it, the idea of objective luck is an illegitimate illusion. At most and at best, there is room only for subjective luck rooted in man's incomplete and imperfect knowledge of how things happen in the world. Luck is no more than a misimpression rooted in human cognitive imperfection and ignorance.

In the wake of the prominent role allotted by modern science to choice and chance in the world's scheme of things—the stochastic operation of physical nature and the drastic complexities of brain processes—such a position is difficult to maintain. In the world order replete with probabilistic principles, denying the role of scheme and haphazard is no easy prospect.

* * *

Perhaps a day will come when the idea of luck can be abandoned with chance and contingency unmoved from human affairs, with Bishop Butler's dictum that probability is the guide of life no longer in order. But from the vantage point of present indications, such a prospect looks to be extremely remote. The condition of mankind being as it to all appearances is in this uncertain world, luck cannot be eliminated as a key factor of our existence, be it in cognitive, practical, or ethical contexts.[14]

Notes

3. This discussion extends and develops some ideas in the author's *Luck* (New York: Farrar, Straus, Giroux, 1995; reprinted Pittsburgh: University of Pittsburgh Press, 2002). An earlier version was published in 2014 in a special issue on luck in the journal *Metaphilosophy*.
4. The verb *to luck (out)* means "to turn out well by chance." (We read in Caxton's *Raynar* [148]]: "When it so lucked that we toke an ox or a cowe.")
5. While some of the best philosophical dictionaries are in German, luck as English speakers know it simply does not exist for these reference works because they do not accord recognition to the fact that something rather special and indeed unique is at issue when happiness or misery (*Glück* or *Unglück*) occurs through sheer accident (*Zufall*).
6. The term was coined by Horace Walpole after the fairytale, set in Ceylon, entitled *The Three Princes of Serendip*.
7. A good indication to the subject and its literature is provided in I. M. Church and R. J. Hartman (eds.) *The Routledge Handbook on the Philosophy and Psychology of Luck* (Routledge: New York & London, 2019). There is a large literature on this topic.
8. As the Bibliography at the end of the book indicates, there is a substantial literature on moral luck. The present author is very skeptical about the matter. For if you do the ethically right thing by chance—e.g., restore something to its proper owner by giving it to a random recipient who "just happens" to be the owner, you earn no moral credit for having kept a promise. If you pick the right medicine for someone entirely by accident you get no credit for providing aid. Strictly speaking, there is no such thing as "moral luck."
9. Jacques Monod, Chance and Necessity: Philosophical Issues in Modern Biology (New York: Knopf, 1971), p. 140.
10. George Gaylord Simpson, "The Nonprevalence of Humanoids," *Science*, vol. 143 (1964), pp. 769–775 = Chapter 13 of *This View of Life: The World of an Evolutionist* (New York, 1964), see pp. 773.
11. See Stephen J. Gould, *Wonderful Life* (New York: Norton, 1989).
12. Theoreticians have had to struggle to make this idea precise. A conjuncture is fortuitous if it involves the concurrent realization of events that are produced by chains of causality that operate independently of each other. Accordingly, the fortuitous does not call for an abrogation of causality but only the prospect of its operation along mutually irrelevant pathways. (Compare A. A Carnot, *Considérations sur la marche des idées et des événements dans les temps modernes in Oeuvres complètes* ed. by J. Mentré [Paris, 1879], Vol. I. pp. 1-15.) A random or

stochastic event, which operates independently of any and all causality, is thereby *a fortiori* fortuitous.

13. On this issue, see the author's *Predicting the Future* (Albany: State University of New York Press, 1998).
14. Philosophers have in recent years extensively debated issues of moral and epistemic and even theological luck. The issues arrive from situations of unintended or inadvertent right- or wrong-doing. When someone acts it right (or wrong) by sheer chance and inadvertence, do the otherwise standard norms apply? In shooting the sleeper whom someone has stabbed to death earlier on, does one deserve moral reprehension for an act of expected and intended murder, when the actual act is no worse than firing a bullet into a corps? In their discrepancy between outcome and expectation, such situations are lucky. But there remains the question of the extent to which such luck affects moral or epistemic assessment and the extent to which outcomes that are good or bad outcomes should be adjudged as such when reached by accident and inadvertence. The reader who takes interest in this issue can consult a large literature, much of it listed in the Bibliography below. However, while the proper question is that of the *effect* of moral or epistemic luck, some of the literature mis-frames it as that of its *existence*.

Introduction

Setting the Stage for Chancy Luck

Luck is a theme that reverberates throughout a wide area of philosophical concern. Ethics will deliberate about the bearing of luck or fairness ("Is it fair that *X* is lucky and *Y* not?"). Metaphysics will ask for explanations of the occurrence of luck and of its reason and range. Political theory will consider the implications of luck for the distribution of wealth and power. But all of this presupposes a prior analysis of what luck is, how it works, and how its presence and its extent are to be determined and quantified. This issue—the mathematics of luck—is the subject of the present book.

Luck is not a power, force, or agency of some sort; it is not a productive potency, but a retrospective explanatory default to which we resort when all else fails. In effect, luck is the unexplained residue of causal accounting. To say that a result is realized "by luck" is to confess ignorance about detail.

As we standardly deal with it, luck is an artifact of insufficient information. It is not an agency of Nature that can be purported and cajoled. It is, instead, a reflection of the finitude of man inherent in the fact that—for whatever reason, be it nature's fecklessness or our own ignorance—we are unable to foresee the outcome of situations in which we have a stake.

Luck hinges on an unforeseeable development impacting positively or negatively upon someone's interests—an agent or group that has a stake in the matter. Benefit and detriment for people by mere chance and happenstance is the crux of luck. And as such it is of course a matter of amount, of more or less, with one selfsame development potentially lucky for some or unlucky for others. It is this matter of measuring luck and elucidating its consequences that is the force of the present book.

Luck (be it good or bad) has a dual aspect. It can figure in relation to *how a result is obtained* (e.g., by luck in contrast with such factors as skill, dedication, or cheating). And it can be *a feature of that result itself* via the negativities or possibilities it brings to those concerned (e.g., "he was lucky that the coin came up heads that gave him the advantage of the opening move in the game"). The former sort of luck is *productive* and relates to how an outcome is realized. The later is *consequential* and depends on the good (and bad) results of that outcome's realization.

The main concern here is with the second issue and primarily with assessing the extent to which an outcome is lucky (or unlucky) in its bearing in the interests of the beneficiary (or maleficiary) concerned in a chancy outcome situation.

It is never a good thing to *need* luck but it is certainly a good thing to *have* it. Retrospectively good luck is splendid; prospectively it is a challenge. And so, like the Roman god Janus, luck is two faced. But it presents a very different visage in these two directions. Toward the past, it is all smiles. When we have been lucky, we have good reason for being pleased. But in lucky toward the future, the situation is very different. For when luck is something we do not yet have but stand in need of, we are far from fortunate. To enjoy luck's favors is a great good but to stand in need of this boon and be dependent upon having good luck is an unhappy condition.

Just this renders the appraisal and measurement of luck a critical consideration. For in relation to the past, it is an index of appropriate gratitude and in relation to the future a monatory warning to caution.

* * *

Luck is a matter of respects: lucky in this, unlucky in that (lucky in business, unlucky in love). The range of luck is virtually unlimited. It can function in every region of human concern: in practical affairs, luck in business dealings (right place at right time) in risk taking; in epistemic/cognitive affairs (lucky/shrewd guesses), successful conjectures; and even in moral affairs (doing the right thing by inadvertence rather than principle).

* * *

Two very different questions arise in deliberations about luck:

- When an agent *X* brought the outcome *O* about, did he do so by luck—or by skill by cheating, or in some other specifiable way?
- To what extent would it be lucky for *X* if outcome *O* were to come about? In the circumstances, could someone reasonably consider themselves very lucky or unlucky?

The former issue is historical, productive, and factual. The latter is hypothetical, evaluative, and analytic. And it is the second issue in specific that will primarily concern us here—the theoretical assessment and quantification of luck. Very different matters are concerned in those two questions.

Occurrences are not just lucky or unlucky in and of themselves; we have to ask: for whom. The issue of beneficiaries and victims is crucial.

Luck is not an unmixed blessing. It has different aspects and plays by different rules. It is never a good thing to *need* luck but it is certainly a good thing to *have* it. Retrospectively good luck is splendid; prospectively it is a challenge.

There was a gas leak and the building exploded. It happened at night when only the night watchman was on the premises who regrettably lost his life. So, fortunately it did not happen in the daytime when 200 people were usually. That was pretty lucky! But for whom?

- It was certainly unlucky for the night watchman, who lost his life in a small-probability event.
- It was very lucky for all those who worked in or otherwise had business in the building at the time. Their lives would have been substantially at risk.
- It was also lucky for the society, which would otherwise have sustained the loss of many productive people in a small-probability event.

Clearly whether or not an event is lucky depends on the "for whom" with respect to those who have a stake in the matter.

Luck prevails when an eventuation that is of benefit or detriment to someone occurs *by chance*—by unforeseen haphazard. "Being lucky" can thus be bound to characterize both people and occurrences: a lucky occurrence being one that is lucky for someone, and a lucky person being someone for whom a beneficial eventuation occurs by chance.

Luck pivots on the idea of chance, duly implemented in our understanding or the world's phenomena in such conceptions as randomness, stochastic phenomena, haphazard, coincidence arbitrariness, contingency, and the like. Its philosophical roots go back to the Aristotelian concept of *tuchê* and the "swerve" of Epicurean cosmology. Luck is to be understood on this basis as a matter of having eventuations that occur "by chance" and affect our interests (be it for better or for worse).

And in this context, chance is not a matter of the absence of consistency or of explainability, but rather one of predictability. Even if a Laplacean determinism through nature's laws pervaded the universe, our inevitable ignorance about detail would leave ample space for chance, randomness, and coincidence of being at the right (wrong) place at the right (wrong) time.

While luck is typically a matter of having things go right (or fail to go wrong) unforeseeably "by chance," it need not necessarily be "against the odds." Sometimes people are lucky even when the odds are on their side. The householder was lucky his place survived unharmed when 40 percent of his neighbors saw their houses damaged by the hurricane. Jones played Russian roulette and lived to tell the tale. He was certainly lucky—even though only one of the six chambers of his revolver was loaded so that the probabilities favored survival—for it was only "by chance" that things turned out well. Again, someone who escapes unharmed in a serious accident is lucky even if this occurred in circumstances where most managed to survive (so that survival was likely).

Luck is not a force, factor, or agency; it makes no sense to appeal to luck for aid. When one is lucky, it is fine to be happy about it, but makes no sense to feel thankful toward it. For at bottom luck is nothing more than a matter of chance, and chance does no one any favors.

* * *

Overall, there are four levels of success attainment in matters of goal realization:

(1) *Luck*: pure randomness (the typing monkeys)

(2) *Amateurishness*: as exhibited in the fortunate success of the inexperienced unskilled and untrained
(3) *Skill and competence*: the capacity to explicit timing and experience
(4) *Expertise*: functioning at the highest level of performative capability

Here the successive enhancements in ability can be measured against luck: with randomness in the baseline.

Finally, it must be stressed that "winning by luck" and "being lucky to win" are very different issues. Different matters are at issue with these locutions. "Winning by luck" relates to how a result was realized as a matter of *process*—viz., by chance and not (say) by contrivance or cheating. By contrast, "being lucky" in achieving a result relates to its consequence as a matter of *product*. When a yield is substantial, one is very lucky, while if only a miniscule amount is at stake, then it makes little sense to claim being lucky to win even when that success is realized by luck.

* * *

In chancy situations, we find ourselves (figuratively speaking) playing a game against Reality in determining outcomes. In such conditions, no matter which way we turn, an inscrutable Reality can respond in ways unforeseeable to us, either because Reality is irresolute and the matter is as yet undecided by it, or because we lack the information needed to determine its resolutions. (It is not just the absence of control by us, but the absence of foreseeability that is the crux of luck.) It is in such circumstances, where we have a stake hanging on outcomes hidden in a fog of unknowing, that the prospect of luck comes to the fore.

In particular, luck—be it good or bad—is not something one can earn, let alone deserve. In fact, it can come altogether uninvited—without one's actually doing *anything* about it. Thus, you may win at cards because your opponent misguesses your having a certain critical card. His misguess is then your good luck although you did absolutely nothing about it. And even Judaic teaching, which generally discouraged gambling for personal gain, tolerated it when some of the winnings went to charity, and was prepared to see the hand of God in such realization holding that he who wins at a lottery should offer the blessing Shehecheyanu.[15] In general, even those who were skeptical about chance were inclined to leave some room for unforeseeability and luck.

* * *

Luck does not necessarily require real physical chance or natural indeterminacy in the events at issue but only unknowing on the part of its beneficiary. Even though it is not in itself chancy or uncertain what awaits in the room behind the door, if you are right that it is the lady rather than the tiger, you are nevertheless lucky. Since there is cognitive as well as stochastic luck, the crux in the matter is not randomness as such, but unknowing.

Despite frequent assertion to the contrary, lack of agent control is required by luck.[16] For one thing, the agent may exercise that control via by random processes: nothing prevents the use of dice. For another he may—like a drunken pilot—actually have control and yet exercise it by unwitting inadvertence. Granted, being lucky is incompatible with controlling deliberately and intelligently. But then it is incompatible with a long list of other things as well—inevitably for one thing. The fact is that lack of control is neither a sufficient nor a necessary condition for luck. You completely and totally lack control over what day of the week tomorrow will be, but its being Tuesday is certainly not something that happens by luck. And when a fledgling pilot successfully lands on a carrier in a pitching sea, he controls the landing and yet is lucky when it comes off well. Control does not necessarily either enjoin or preclude luck; sometimes it involves luck, sometimes it excludes it.

Moreover, favorable outcomes which are entirely beyond an agent's control may or may not be matters of luck. Thus, winning the lottery could certainly be so, but falling heir to your parents' estate would not. (People may not control nor even foresee what their parents leave to them in their will, but receiving such a bequest is surely not a matter of luck.) Again, you do not control whether it is summer or winter but neither way are you lucky (or unlucky) about it. When you stop at a traffic light, you have no control over it. But when it turns to green and speeds you on your way at a juncture you neither controlled nor foresaw, there is nothing of luck about it. Altogether, it is more conducive to clarity to conceptualize luck in terms of chance rather than agent control or its lack.

Notes

15. *Encyclopedia Judaica*, Vol. 7, p. 302a.
16. See, for example, Zimmerman 2002, p. 559. (References of this format refer to the Bibliography at the end of the book.)

Contents

Notational Conventions

Glossary

Chancy situation (stochastic outcome situation or *situation* for short): a circumstance in which various alternative outcomes are possible by which someone stands to gain or lose.

Circumstance: a condition in which chancy situations arise.

Outcome: the possible result of a chancy situation.

Agent or *protagonist*: the party who stands to benefit or disbenefit from the outcome of chancy situations.

Beneficial/detrimental: condition of outcomes correlative with the agent's gain or loss.

Lucky/unlucky: evaluative polarity (as positive or negative) of the luck status of an outcome.

Risk: the prospect of a negative outcome; the opposite of *promise*, the prospect of a positive outcome.

Notation

A modest amount of formal machinery is needed for the mathematically informed discussion of luck.

O_1, O_2, O_3,. . . alternative outcomes constituting the *range of possibility* for outcomes in a situation of uncertainty.

$pr\{O_i\}$ or p_i for short: the probability of realizing alternative outcome O_i.

$|O|$: the subject's *yield* by way of gain or loss (benefit or disbenefit) resulting from the realization of outcome O.

$|O| \times pr\{O\}$: the yield-potential *moment* of outcome O.

O^+: the *optimal outcome*, so that $|O^+| \geq |O_i|$ for all i.

O^-: the *minimal (or "best possible") outcome*, so that $|O^-| \leq |O_i|$ for all i. (Thus, $[O^-]$ is, so to speak, the "worst-possible" result. It represents the *gross risk.*)

Δ: the size of the range that defines the *stake* at issue in the chancy situation at issue. (Thus, $\Delta = |O^+|^* - |O^-|$ when $|O^+|^*$ is the ___ $|O^+|$ and $|O^-|^*$ the universal O^+ over the range to alternative outcome.)

$\lambda\{O\}$: the *amount of luck* (positive or negative) at issue with the realization of outcome O (or required for it).

$\lambda^-\{O\}$: the *degree or comparative extent of luck* afforded a chancy outcome. An outcome is λ^--lucky whenever $\lambda^-\{O\} > \frac{1}{2}$. Note: The two measures of luck λ and λ^- agree about rank-ordering the outcomes.

$\mathcal{L}\{O\}$: the *luck differential* of an outcome with $\mathcal{L}\{O\} = \lambda\{O\} - \lambda\{\text{not - }O\}$.

E: the *expected value* of the yield in a chancy multiple-outcome situation: the sum total for all outcomes of their various yield-potential moments.

E^+: the *positive part* of the expectation E of a chancy situation encompassing all (but only) the positive-yield outcomes (analogously with E^- for the rest).

θ^+: the positive luck potential of a chancy situation, namely, the total, overall probability of realizing a positive outcome yield. Thus, $\theta =$ the sum total of all p_i of these outcome O_i whose yield is positive: $|O_i| > 0$. (Note: $\theta^- = 1 - \theta^+$)

Basic Equations

- $E = \sum_i (p_i \times |O_i|)$

  E{S} or simply E, the sum total of all the yield moments of the totality of outcomes of chancy situation S_n. This measures the *expectation* of that chancy situation.

- $\lambda\{O\} = Y\{O\} - E = |O| - E$

  λ represents the *amount of luck* of the outcome at issue. [Note: This can range from $-\infty$ to $-\infty$.]

- $\lambda^-\{O\} = \dfrac{\lambda\{O\} - \lambda\{O^-\}}{\lambda\{O^+\} - \lambda\{O^-\})}$
  $= \dfrac{(|O| - |O^-|)}{|O^+| - |O^-|} = \dfrac{|O| - |O^-|}{\Delta}$

  λ* represents *the loss avoidance luck* of the outcome at issue. [Note that λ^- measures a *degree of luck* which ranges from 0 to 1.]

- $\lambda^+\{O\} = \dfrac{|O^+| - |O|}{|O^+| - |O^-|}$
  $= \dfrac{|O^+| - |O|}{\Delta}$
  $= \dfrac{\lambda\{O^+\} - \lambda\{O\}}{\Delta}$

  λ^+ represents *opportunistic luck.* [Note this is *a degree of luck* which ranges from *O*+ to ∞.]

- $\lambda^\#\{O\} = |O| - E^+$

  $\lambda^\#$ represents *second-order luck* indicating how a lucky outcome stands in the overall range of such outcomes. [Note that $\lambda^\#$ goes from $-E^+$ at $|O| = 0$ to $+\infty$.]

- $\mathcal{L}\{O\} = \frac{1}{1-p} \times \lambda\{O\}$

  $\mathcal{L}\{O\}$ is outcome *O*'s *luck differential*, as per $\lambda\{O\} - \lambda\{\text{not-O}\}$.

Chapter 1
Chancy Outcome Situations

The common expression speaks not improperly of "dumb luck," for what happens to people "by luck" cannot be foreseen by them. One may be fortunate through the realization of a securely predictable boon, but (by hypothesis) it is not going to be a fruit of mere luck. The role of chance in the world has been deliberated philosophers since antiquity. Thus Aristotle considered whether chance (*tuchê*) and accident (*to automaton*) are fortuitous or causally explainable—whether they have standard, causal explanations (as he understands this) or themselves constitute a causality of sorts (*Physics,* II, iv; 195b30ff). He concluded (1) that it is reasonable to think that whatever is said to happen by chance actually has a causal explanation but that owing to the complexity of world processuality and the role of chaos and chance the causality of chance it may well be unexplainable to the human intellect, (196b6-7), because that causality, though real, simply being unintelligible. Were this so, chance and luck would ultimately be artifacts of ignorance—of the inevitable limitations of finite intellects. This position is not far from what still deserves to be seen as plausible.

A chancy *outcome situation* prevails in human affairs in circumstances where there are various possible alternative outcomes subject to haphazard or chance. They may be binary in having just two possible outcomes (as with reacting to a proposal of marriage). Or there may be multiple, as with deciding on a car purchase. The outcomes of such situations can obviously vary and yield different results.

One can have a stake in a chancy situation without knowing about it. (Think here of the potential victim of an aborted assassination plot or of the baby whose parents have bought a winning lottery ticket for its "college fund.") Accordingly, one can be lucky (or unlucky) without being aware of it. And a qualified theory of luck can go only as far as its prerequisite theory of value enable it to do.

* * *

In situations of unexpectable outcome there are often alternative ways of facilitating the realization of a favorable result. Both skill and cheating are notably

N. Rescher, *Luck Theory*, Logic, Argumentation & Reasoning 20,
https://doi.org/10.1007/978-3-030-63780-4_1

prominent among these alternatives. But in those circumstances where chance is dominant, the matter is one of luck.

Luck pivots on inadvertence and impredictability: results realized in ways other than by chance—by skill, effort, persistence, planning or the like—are not matters of luck. When an outcome's secure predictor would require information that is not—and very possibly could not—be available, then its ascription to chance is in order. Such foreseeability can arise in a variety of ways. For an occurrence can be

- in fact unforeseen by the specific protagonist at issue
- in principle unforeseeable by this individual on the basis of the information in his possession
- in principle unforeseeable by this individual on the basis of the information assessable to him/her
- in principle unforeseeable by this individual on the basis of then-available accessible information at his disposal.

All of this gives rise to a mode of chance and thereby opens the door to luck. And on this basis the individual who correctly answers a questions thorough guesswork or conjecture qualifies to be considered lucky about it.

However, chance is not something that lies "in the eyes of the beholder." It can be an objective matter, like *being* healthy, rather than a subjective matter, like *feeling* healthy. For a situation is chancy for someone when foreseeability prevails because it is infeasible, given the information at this agent's disposal, to determine which possible outcome is to eventuate. And this is in general a factual matter of general information rather than of a personal disability of some sort.

* * *

Luck Theory is part of Rational Decision Theory and accordingly requires two quantitative factors for its operations: values and likelihoods—the utilities and the probabilities of outcomes. And specifically, the Calculus of Luck needs to have at its disposal both a Calculus of Probability and a Calculus of Value. Yet neither is altogether straightforward and unproblematic.

First take outcome probability. This comes in two versions: *subjective* or *personal* probability as assessed by an individual's own (possibly erroneous) thinking about the matter, and *objective* or *impersonal* probability, based on considerations of statistical frequencies of occurrence or physical symmetries of constitution. This distinction between objective (statistically measurable) probability and subjective (personally appraised) probability is crucially important for an understanding of luck which itself can bear either of these two aspects. And only when the probabilities at issue are assessed cogently will one be dealing with luck in a sensible manner.[1]

Second as to evaluation. This also comes in two versions: *subjective* in depending on personal assessments, and *impersonal* in being determined by the impersonal factors such as marketplace in economics.

One must accordingly distinguish between impersonal luck based on measurable quantities and subjective luck based on personal sentiments. All of the factors at issue with luck—yield evaluation, outcome probability, and result estimation—can be assessed both on a personally subjective basis or with reference to measurably objective factors. Consequently, there is both subjective and objective luck. However, the present deliberation will focus on the objective and measurable dimension. Formal Luck theory belongs to the theory of cognition not to personal psychology. The subjective psychology of luck is an important but separate issue.

And so throughout these discussions careful usage must accordingly distinguish between *the* (objective) yield, *the* (objective) probability, and *the* (objective) luck of an outcome, and *your* (subjective) yield, *your* (subjective) probability, and *your* (subjective) luck and for present purposes this distinction is pressingly necessary because our aim here is throughout to deal with the objective side of the matter, so objectivity will be presumed. It is the mathematics and not the psychology of luck that concerns us here. A quantified theory of luck can carry us only as far as its prerequisite theory of value enables it to do.

* * *

Specifying the probability of an event is all too often not something straightforward. For one thing it depends on evidence and information. Freezing water into solid ice sounds far-fetched to people living in the tropics; reliable sunshine sounds unlikely to inhabitants of the Shetlands. Probability is not always a tractable and well-defined factor: it can be indefinite and problematic. There is no determinable probability that Judge Crater was alive in 1950 or that Lizzie Borden killed her parents. Crude approximation is often the best one can do. The substantiation of probability specifications is a complex business involving abstraction, statistic, scientific reasoning, and plausible conjecture. And probability evaluation has many involvements in applied mathematics, decision theory, economics, inductive inference, etc. In this regard, luck theorists sail in the same boat as many others.

And parallel observations, of course will apply in respect to the evaluation of yields and the estimation of probabilities. Probabilities and utilities (values) alike have an objective and a subjective dimension. And this distinction will carry over to luck as well.

However, it makes no difference for *the process of calculation* of risk or luck whether the salient quantities at issue—the yields and probabilities—are objectively meaningful or somehow personalized and subjective. Here as elsewhere the matter of how a given individual *does* view something and how that hypothetical idealization "the rational individual" *would and should* look at is poses different issues and address different questions. Obviously in matters of personal planning or biographical explanation the subjective domain is at the forefront. But with issues of such deliberations and public policy these issues of risk and outcome have to be treated in the most objective possible way. As indicated for present purposes probabilities, outcomes, and expectations are to be construed in the objective mode, with personal (psychological) probabilities and other subjective evaluations reserved for the psychological treatment of luck-related issues.

* * *

Being lucky and "achieving one's aim's by luck" are not quite the same thing. If someone returns your lost wallet you are indeed lucky. But you received it not by luck but through the kindness of strangers. If you make every flight connection of your complex itinerary on time, you are lucky. But this good outcome resulted not by luck but through elaborate efforts by the airlines. Even outcomes that cannot be ascribed to chance can prove lucky for their beneficiaries—provided they are in no secure position to foresee them. However, access to information is a crucial factor. When his friends rig the lottery in *X*'s favor he has every right to think himself lucky, even though they "know better" as it were.

Recall Frank R. Stockton's classic story of *The Lady and the Tiger*.[2] No matter now firmly the forces of the universe have conspired to put the maiden behind the one door and the tiger behind the other, if our protagonist makes the right choice by mere guesswork he is in fact very lucky. The total absence of relevant information renders the matter one of totally chancy guesswork for him.

* * *

Pained by a stomachache, Smith choses the wrong medicament, one that worsened matters. Was this unlucky? It was certainty unfortunate. But whether it was unlucky or not depends crucially on the explanations of why he did it.

- If his masochistic intent was to enhance his suffering, he may have been stupid, but was not unlucky.
- If he acted in response to erroneous information, he was unfortunately uninformed, but not unlucky.
- If he acted by mischance and mistakenly picked up the wrong bottle in simple confusion, then he was clearly unlucky.

It is the essential role of chance that qualifies this third case as a matter of luck, given the fact that what was operative here was not perversity or misunderstanding or some other such causally explanation factor, but mischance, haphazard, and fortuitous inadvertence.

* * *

Expectation of outcome-yield in chancy situations is a critical factor with luck. To be sure, we sometimes assign great luck to the realization of modest benefit. Thus the person who finds the house key he lost in crossing a large field may be deemed to be very lucky. (The key is inexpensive but the inconvenience of its replacement great.)

We shall represent the various outcomes issuing from the branching points of a chancy situation by letters O_1, O_2, O_3, ... , and will use the letters *p*, *q*, *r*, etc. to represent the probabilities of their realization. The yields for the party at issue occasioned by the outcome O_i (be it a gain or loss) will be represented by $|O_i|$. Thus the overall chancy situation may be represented graphically by an "occurrence diagram" of the format illustrated by:

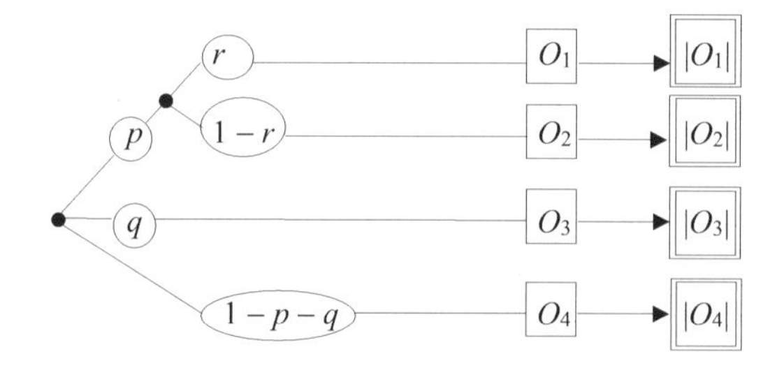

Reading from left to right in here, there is first a branch-point pf resolution from which there issues, with variant probabilities, one or another of several possible outcomes O_i. These in turn have a certain specifiable yield $|O_i|$ for the party whose interests are at stake. And where there are difficulties in the assessment of outcome probabilities or yields corresponding difficulties in the assessment of luck can become unavoidable.

To be sure, sometimes branching is the product of decision so that a branching point becomes the index of the agent's option (then distinguished by □ in place of •). Then if the agent is the situation's protagonist, then there is nothing probabilistic to it. But when the choice is someone else's, then probabilities enter in once more. (For me your decision is just another chancy outcome.)

With gain and loss alike, inevitability will exclude luck. There is no room for luck when chancy situations are yield-uniform across the whole spectrum of possible outcomes. Otherwise, however, there will always be some lucky and some unlucky outcomes. And whatever outcome-value is maximal will ipso facto maximize luck: one cannot ask for more than the best-available possibility, and one is lucky to achieve it.

A chance situation is *risky* when its outcomes differ in point of yield, so that there is a significant distance between the best and worst outcome—between $|O^+|$ and $|O^-|$. Again a chancy situation will be *unfortunate* in those unhappy situations where every possible outcome is negative, so that not just $|O^-|$ but even $|O^+|$ represents a loss. Finally, a chancy situation is *noxious* when its worst outcome yield $|O^-|$ is at once both highly negative and non-trivially probable. Paradoxical though it seems, a chancy situation can be noxious and yet involve little risk or bad luck. This will transpire in the exceedingly unhappy situation where every possible outcome is negative so that a bad outcome is inevitable. (One may be unlucky to have got into such a situation, but once there, there is no further room for luck.)

* * *

We now turn from evaluating the outcome-yields of chancy situations, to evaluating the comparative acceptability of those alternative chancy situations themselves. The salient question here is: just how favorable or unfavorable one situation is in comparison with others. What are its comparative prospects for yielding a favorable overall result?

It is essential for a theory of rational decision to be able to assess the overall value of chancy situations as such because our options in chancy matters are all too often themselves mere chancy situations. And then, instead of yielding a specific result, we arrive at a further chancy situation. (For instance, rather than resulting in an assured gain of 3 apples, the outcome-yield of a chancy situation may be a 50 percent chance of gaining 5 apples—or nothing). Such issues will also have to be reckoned with in an uncertain world where the result of our choices are all too often "up in the air" with regard to alternatives.

* * *

At this point rational decision theory returns to its origin in the days of Pascal and Leibniz in the 17th century and adopts the conception of "expectation" (Latin *spes*), with its characteristic amalgamation of the values and probabilities of possible outcomes.

Thus consider a chancy situation with an outcome manifold $O_1, O_2, O_3, \ldots$, each member with its own particular value $|O_i|$ and its respective probability $pr\{O_i\}$. Multiplied together these provide a yield-potential *moment* $|O_i| \times pr\{O_i\}$ for each outcome. And the overall "expectation" of the chancy situation is then simply the sum total of these outcome-correlative quantities.

With the yield-potential *moment* of a possible outcome O is its probability-qualified yield $|O| \times pr\{O\}$.[3] And critical for the assessment of luck is the idea of the *expectation* (or "*expected value*") of the chancy situation for the party at issue, as constituted by the total *yield-potential* of its various outcomes—*the probabilistically weighted average of the outcome-yields that are involved*:, as defined by:

$$E = \sum_{i} (p_i \times |O_i|)$$

The mathematical expectation E of a chancy situation is in effect a more sophisticated, "weighted" average of its outcome yields—one that takes account of the fact that their realization is not equiprobable.

Duly defined for the overall situation, this quantity in effect provides a probabilistically weighted averaging out of the potential outcome-yields of the various possibilities. And as such it renders possible the assignment of one definite overall quality to represent an otherwise complex and varied uncertain-outcome manifold.

Three factors are thus of prime significance with expectations: (1) the range of alternatives (2) the results that those alternative outcomes yield by way of benefit or cost, and (3) the respective probabilities of their realization.

The idea of "expectation" needs further commentary, having both an informal and a technical sense. The latter is readily explained.

One significant evaluative factor bearing is the prospective *moment* of an outcome O as defined by the product of its yield $|O|$ and its probability: $pr\{O\} : |O| \times pr\{O\}$. The sum of these outcome moments for the totality of possible for all outcomes constitutes the overall *expectation, the probability-weighted average of the possible-outcome values:*$E = p_1 \times |O_1| + p_2 \times |O_2| + p_3 \times |O_3| + \ldots$ This approach to expectation management, does useful and instructive service throughout the whole range of matters of rational evaluation and decision. It provides an indispensable resource for the present theory of luck assessment, because "expectation" in just this manner is a statistically cogent estimate of what will happen in the long run when the relevant probabilities are in play.[4]

* * *

To see how expectations work, consider the situation of a die-toss yielding a win of as many dollars as the die shows points. Then that 6-membered range of possible outcomes would yield an expectation of

$$\frac{1}{6}\times 1+\frac{1}{6}\times 2+\frac{1}{6}\times 3+\frac{1}{6}\times 4+\frac{1}{6}\times 5+\frac{1}{6}\times 6=\frac{21}{6}=3.5$$

So \$3.50 would in these circumstances be the overall expectation of this chancy outcome situation, its probabilistically weighted average yield. And this provides our best-available assessment of the overall yield potential of the situation.

The example of tossing a coin subject to someone's gain of \$1 for Heads (H) and loss of -\$1 for Tails (T) provides for the following chancy-outcome situation:

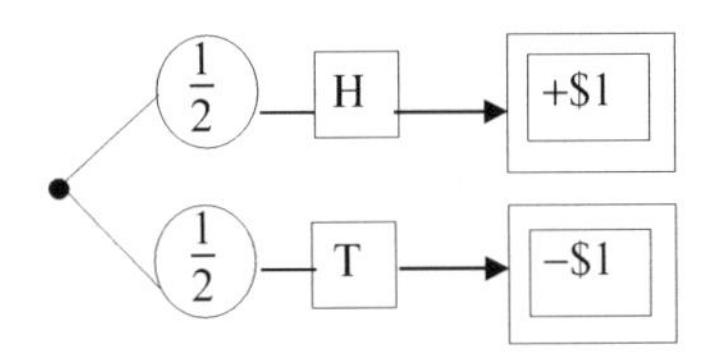

The expected value of this situation will accordingly be:

$$E=\left(\frac{1}{2}\times +1\right)+\left(\frac{1}{2}\times -1\right)=0$$

The zero expectation prevailing here means that the gamble involved is neither gain-favorable nor loss-favorable but "fair," that is, equally balanced between the possibilities. Given the nature of probabilities, it is clear that such an expectation provides useful guidance in estimating the potential yield of such a chancy situation.

* * *

On this expected-value approach to the yield-prospect of chancy situations, it transpires that a low-probability high-loss outcome as per

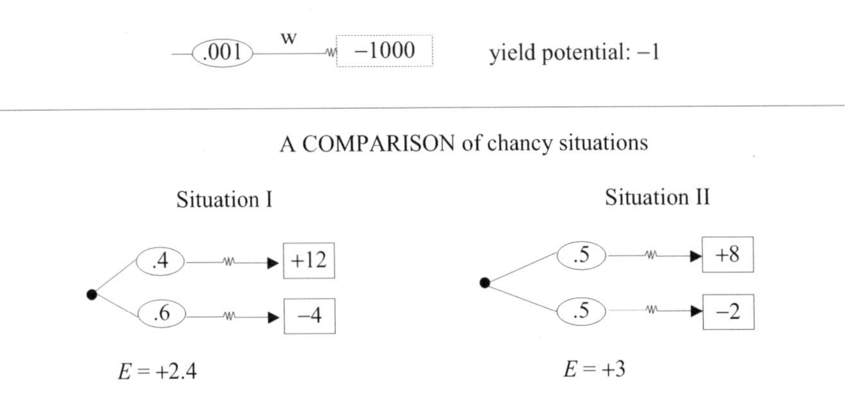

is equivalent in its estimated value with a high-probability, low-loss situation as per:

.1 —w— −10 yield potential: −1

As the concept of expectation is defined, each makes the same contribution to the overall total.

* * *

Given the nature of the probabilities involved in such a calculation this quantity represents the average result that would be expected for a large number of repeated iterations of the chancy situation at issue.[5]

Its overall expectation affords a rough (approximation) acceptability-evaluation for a chancy situation: a positive expectation indicates a favorable prospect, a negative expectations an unfavorable one. A zero expectation such as that of the previous example, indicates that the situation constitutes what is termed a *fair* gamble, namely one where the prospects of winning and losing are equally balanced equal for the parties invoked. (In actual gambling practice, the house is a profit-seeking business and never offers fair gambles.)

* * *

Again, if the chancy situation at hand were the winner-take-all one dollar gamble of a head-or-tail coin-toss we would have

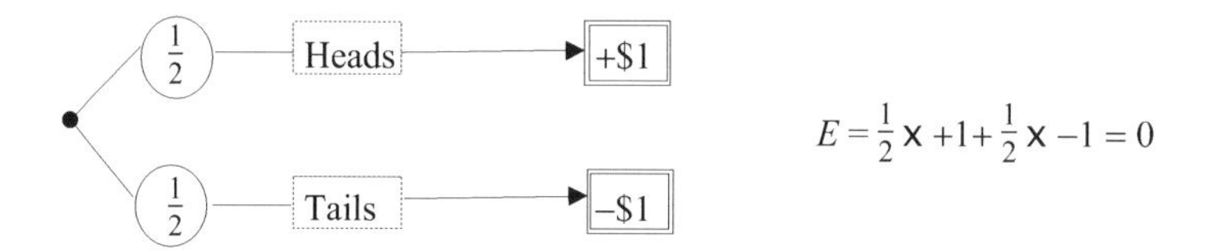

Its zero expectation marks the situation as an even proposition, a fair gamble. In the long run those taking the gamble should break even.

* * *

Sometimes one is embarked on a processual pathway leading from one chancy situation to another. Such a proceeding might, for example, have the format:

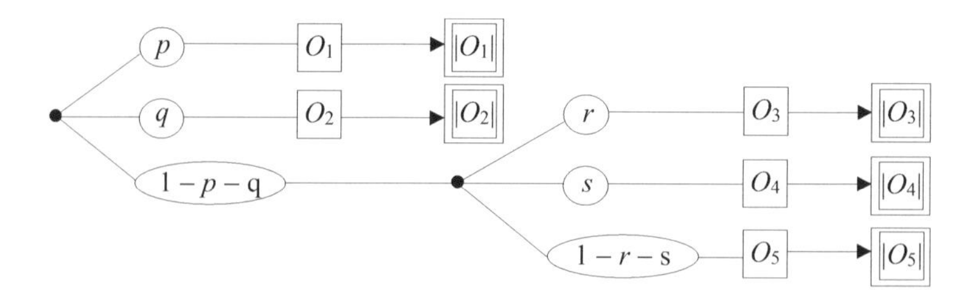

The following reformulation shows how such a multi-stage situation can be recast into a single stage equivalent, as per:

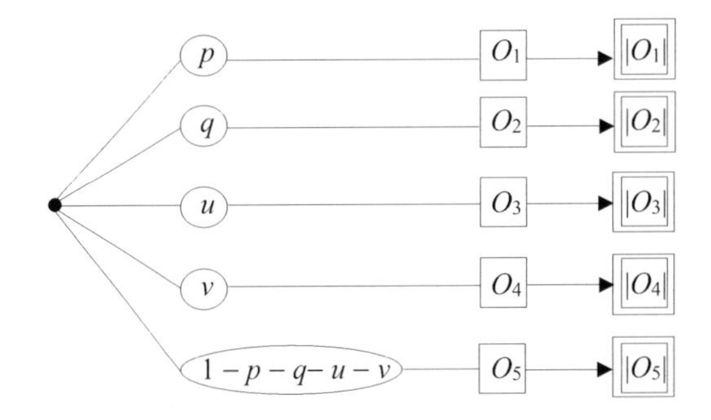

To be sure, a residual problem here is to calculate *u* and *v* in terms of the original probabilities *p*, *q*, *r*, *s*. This can be achieved via the equations:

$$(1 - p - q) \times r = u$$

$$(1 - p - q) \times s = v$$

The idea here is that the outcome yield of a chancy situation can be assessed as the amount of its expectation.

As this illustrates, it is always possible to contrive a single-stage representation of multistage stochastic situation.

Notes

1. For a clear discussion of expectation claims in this regard see Brian Skyrms, *Choice and Chance* (Encino, CA: Dickenson Publishing, 1975; 2nd ed.), pp. 149–153.
2. This bit of fiction, first published in *The Century* magazine in 1892, has since been anthologized times without number.
3. A further complication is that of *continuous outcome functions* in place of the discrete alternatives with which we have been dealing. This change from a sequence of discrete outcomes O_1, $O_2, \ldots O_n$ to a continuous outcome factor $O(x)$ calls for a change in representational machinery by replacing numerical valuations with continuous functions.

   On this basis treat all of the issues here addressed via arithmetically discrete applications discrete can also be addressed analytically and continuously via the calculus. But while this would substantially expand and extend our range of consideration, it would add nothing fundamental to the picture of luck comportment that already emerges at the arithmetically discrete level.
4. On expectation see the statistical reference provided in the Bibliography as the end of the book, in particular, Fella 1950.
5. On the underlying "Law of Large Numbers" see any treatise on the Calculus of Probability (for example Feller 1950).

Chapter 2
Outcome Luck Assessment and the Luck Equation

2.1 Preliminaries

Luck—be it good or bad—enters when an uncertain, unforeseen outcome is beneficial (or detrimental) to someone. Its assessment is a matter of the extent to which unforeseeable developments impinge on people's interests. The calculus of luck fuses the calculus of probability with value theory to provide a means for assessing the benefit (or detriment) that chancy evaluations provide for people. Any adequate account of the matter must determine not only whether and how it is that an agent is lucky or unlucky but must also enable us to tell to what extent this is so. It must, in sum, be quantitative. And the Calculus of Luck is the project of measuring extent to which the outcome of a chance eventuation represents good (or bad) luck for the individuals involved.

Chance is the homeland of luck. A person is lucky (or unlucky) when in a chancy situation an uncertain and unexpectable outcome redounds to their benefit (or loss.) Luck dwells in the gap between the actual outcome for someone and their expectation. When outcomes eventual to your advantage and do so beyond your expectation, you are lucky, when they fall short you are unlucky.

Attributions of luck have to be kept within limited horizons. Consider: You win the lottery. (What could be a better piece of luck than that!) But your new-found riches embolden you to take greater and greater risks. So in the end you "Bet the farm" and "Lose your shirt." Even what modest assets you had to begin with are gone! What a tragedy! Yet all the same, initial lottery win stands secure as a piece of luck, nowise unraveled as such by those later misfortunes? Were it otherwise—if luck depended on how things turn out in the end, in the ultimate convolution of things, then we would only be able to make luck allocation posthumously—if then! Insofar as luck as a determinable and quantifiable factor it has to be assessed locally, in relation to specific, near-term situations.

N. Rescher, *Luck Theory*, Logic, Argumentation & Reasoning 20,
https://doi.org/10.1007/978-3-030-63780-4_2

Luck is highly context-dependent and situation-coordinate. Thus suppose you play a game with someone under conditions where you are committed to paying twice your winnings to a third party. Then when you are lucky in relation to your initial opponent, you are correlatively unlucky in regard to your situation with that interloper.

"Can one be lucky when risking a loss?" Of course—namely when all of the other possible outcomes are worse. So even in misfortune one can—oddly—qualify as lucky.

* * *

To speak of "dumb luck" is very much in order, because if an outcome were achieved by identifiable productive factors, such as planning, skill, or smarts, then luck would be out of the picture. When a goal is attained by skill, hard work, persistent effort, or talents, its successful attainment could hardly be ascribed to luck. The successful candidate who wins the election after endless hours of elector-schmoosing, handshaking, and baby-kissing would be rightly vexed at having a victory—however likely or not—credited to luck. When an outcome is realized by design, skill, result-rigging, cheating, or by any such pre-determining ways, luck does not enter in. Only where there is chance, inadvertence, and absence of control or predictability will luck be at issue.[1]

* * *

There is a decided difference between a "lucky outcome" (a result that is for an agent in impacting on his interests) and an "outcome due to luck" (a result achieved by happenstance and chance). The player of a game like tennis or chess can be lucky in winning a match against a superior opponent without necessarily having won a single move or play by luck—i.e., by something other than his best skill.

With being lucky the question "for whom?" is critical. Suppose you ask 10 people to guess whether the coin you are about to toss will come up Heads of Tails. With random guessing, it is virtually certain and only to be expected (with probability almost 1) that the luck of "someone's getting it right" is virtually nil. By contrast, it would be rather lucky for any one the particular group member to get it right.

Since chance and randomness are critical here, luck knows no favorites—and for that matter no enemies. To be sure, we are often inclined to agree with Damon Runyon's report that "I long ago came to the conclusion that all life is 6 to 5 against." But there is something democratically egalitarian about luck, since chance looks with equal indifference upon the good and bad alike.

* * *

Luck pivots on unpredictable outcomes. A chancy, uncertain outcome situation exists whenever for aught one can tell events can issue in any one of several different outcomes. In such stochastic situations we will have various possible outcomes O_i with respective probabilities p_i and yields $|O_i|$.

There are two modes of luck, which may be called, respectively, odds luck and yield luck. The former involves benefit that go "against the odds." Odds luck is thus straightforwardly a matter of probability—the less likely the more lucky the outcome. Yield luck is the difference between the expectation of a chancy situation and its actual yield. Accordingly, yield luck must be assessed in terms of yield-units (dollars gained, time lost, lives saved). Given such diversity, one cannot automatically compare luck across thematically different situations. Resolving such questions as whether one is someone luckier in gaining four more apples than they would be in gaining three more oranges calls for having a common standard for comparative evaluation.

When speaking of luck we generally have yield-luck in view. But in its regard good luck looks very different depending on the direction from which one views it. Retrospectively is a source of pleasure and delight: to have had good luck is a very fine thing. But prospectively the situation is very different. To require luck for reaching one's goal is always worrisome.

These two perspectives give the same result from different angles—both alike, answer the same question: How large is the gap between yield and expectation? For the question, "How much has filled the gap?" and "How much would fill the gap?" come to the same thing. To need good luck to achieve some as-yet-unrealized objective is something ominous and daunting—a ground for worry and dismay rather than satisfaction. So in prospective outlook that outcome is best which relies the least on good luck and minimize it, while in retrospect that outcome is best which affords luck its largest scope in minimalizing it. (With negative yield the reverse is of course the case.)

2.2 The Basic Luck Equation

A metric theory of luck has two indispensable prerequisites:

1. A theory of value for assessing their outcome yields that define the stakes at issue.
2. A theory of probability for assessing the likelihoods of outcomes.

Without quantitative outcome appraisals luck theory lacks the input material required for generating an informative and usable output. In regard to this requirement for probabilities or utilities (as economists all them) luck theory is no different from and no worse off than—personal economics, decision theory, or rational choice theory or any other sector of practical reasoning.

The luck *provided by* the realization of an outcome is the exact counterpart of the luck *required for* its realization. Whenever the realization of an outcome is very lucky for you, then it takes a great deal of luck for that outcome to be realized. So one single measure $\lambda\{O\}$ does service both ways. This circumstance constitutes the Principle of the Conservation of Luck: the luck yielded by an outcome is always equal to the luck needed for its realization: What is to go out must come in.

* * *

Luck can also be assessed by considering at the outcome's position in the range between $|O^-|$ and $|O^+|$. This involves two lines of approach:

- $|O| - |O^-|$: the measure of an outcome's *relief* in averting "the worst possibility," the maximal loss.
- $|O^+| - |O|$: the measure of an outcome's *disappointment* in failing to reach the maximum-yielding optimum

These factors as yet do not however, take the matter of chance and probability into account. Here we need to shift to the analysis:

$|O| - |E|$: a measure of excess over or shortfall from the expectation.

This clearly encompasses both outcome yield and probability.

The yield of an outcome in a chancy situation may be thought of as having two components, conjoining the part provided for by informatively grounded expectation (E) with the part provided for by an expectation-supplementing chance (L). Accordingly we have it that:

$$Y = E + \mathrm{L}$$

The luck of an outcome is determined in relation to its situation's expectations. One is lucky to the extent that achievement exceeds warranted expectations and unlucky to the extent that it falls short.[2] People who realize chancy positivities—winning a bet, say, or realizing a familial inheritance—can be accounted as lucky to an extent that exceeds the average (which, after all, is the best they can reasonably expect).

* * *

Luck, so conceived, is the explanatory residue of available explanation. It is, in effect, a matter of *unexpected benefit*: a measure of the extent to which that which happens exceeds (or falls short) of expectation. Outcomes that exceed reasonable expectations are always lucky to just that extent.

Accordingly, luck's appropriate measure looks to be: [Yield] – [Expectation] or $Y - E$.

Whenever the actually resulting yield exceeds your expectations you are in fact lucky; when it falls short, you are unlucky. Accordingly, the *Basic Luck Equation* for assessing the amount of luck of a particular outcome O for an interested party in a stochastic situation is:

$$\lambda = Y - E \text{ or more fully expressed } \lambda\{O\} = |O| - E$$

with λ for luck, Y for the yield $|O|$ afforded the agent by the outcome O at issue, and E the situation's overall expectation. From a metric standpoint, the amount of luck of an outcome thus comes to the difference been that outcome's yield for an agent and the actual expectation. On this basis, one is lucky to the extent that yield exceeds expectation and lucky to the extent that it falls short.

It warrants note that the negative of an outcome's luck, namely $\lambda\{O\} = -(|O| - E) = E - |O|$, represents what might be called the *disappointment value* of the outcome, its shortfall from expectation.

* * *

Luck cam be viewed from various angles. One pivots on the question: to what extent is *the amount of the outcome-yield* positive (or negative)? Another asks about *the extent to which the outcome averts the worst possible outcome*. (After all, even a negative outcome is to be welcomed when it avoids prospects that are even worse.) But the best plan is to see luck as measurable by *the extent to which an outcome meets (or even exceeds) expectation.* What renders this measure optimal is that it alone takes the crucial issue of likelihood into account. Realizing a good result despite its low odds is the very essence of good luck.

The prospect of good or bad luck is almost ubiquitous throughout chancy situations. For an outcome to qualify as (positively) lucky it must be that $|O| - E > 0$ and so $|O| > E$: its yield must exceed the expectation. But in chancy situations there is virtually always the prospect of a disappointing outcome when $|O| < E$. (The only exception is the extraordinary case where all possible outcomes are equi-valued.)

The comparison that lies at the heart of luck has three possible outcomes:

$Y > E$. In this case, the individual is lucky.
$Y < E$. In this case, the individual is unlucky.
$Y = E$. In this case, luck does not really come into it.

The luck-definitive difference $Y - E$ thus determines both the direction and the magnitude of the luck at issue.

On the basis of $\lambda = Y - E$, the amount of luck involved in a one-chance-in-ten win of $110 [namely $(1 - .1) \times 110 = 99$] is much the same as the amount of luck in wining $1000 in an unrealistically favorable nine-chances-in-ten lottery [namely $(1 - .9) \times 1000 = 100$]. As far as the luck of the matter goes, the greater yield at issue in the second case is offset by the far greater likelihood of its realization. While the yield is more substantial, the high probability of its attainment takes much of the luck out of it.

* * *

The Basic Luck Equation can be regarded from another point of view. Clearly if $\lambda\{O\} = |O| - E$, then $|O| = E + \lambda\{O\}$. On this latter basis the outcome can be seen as compound of the luck with the *expectation* as (based on skill, effort, ability, and the like). Luck so regarded, becomes the gap-filler between evidentially grounded expectation and the actual realization. Thus if you achieve victory in a contest with someone who defeats you a quarter of the time (so that your chance of winning stands at 75%), then when you win one could ascribe 25% of the victory to luck. Your victory was to be expected but still would not be counted upon: to a modest extent you were lucky to bring it off. (Of course not really as lucky as you would have been had your victory come against someone you beat only half the time.)

Consider a situation that is very common in contemporary life—waiting for service (be it in a doctor's office, or a telephone access, or a delayed transport departure). Suppose that the circumstances the normal, expected waiting time is half-an-hour. If your actual outcome is 10 minutes, you are lucky—to the tune of exactly

20 minutes ($Y - E = 10 - 30 = -20$). But if you have to wait an hour your luck extends to $60 - 30 = 30$ minutes. (Note that here, with luck measured in delay-time, more is actually less (i.e., less luck) and less more.)

The units in which this quantity is measured is determined by the sort of value at issue with O: money gained or lost, time spent or saved, contests won or lost, meals eaten or skipped, or some analogous positivity or negativity.

So regarded, luck in these chancy situations is a matter of the excess or shortfall of outcomes with respect to the concerned party's appropriate expectations. And the luck of a possible outcome is always a function of its relationship to the alternatives—it is never something determinate in and of itself.

* * *

In conflict, being attacked by surprise—be it strategic or tactical—is always bad luck. For its outcome is bound to be more negative than otherwise, and its likelihood is in the very nature of the affairs seen as improbable, so that $Y - E$ is one big negativity.

However, the fact that luck is determined in relation to expectation has the somewhat odd-seeming consequence that a negative outcome (a loss) will be lucky whenever a bigger loss is in prospect. It may seem somewhat strange to say that someone was lucky to lose \$1000. However appropriate contextualization should remove the oddity: It sounds perfectly natural to say "He was lucky to lose *only* \$1000 when could easily have lost a million."

Achieving the best-possible outcome in a chancy situation is not always lucky. For when all possible outcomes have the same yield so that $|O|$ is uniformly Z, then the expectation will be $E = Z$, so that $\lambda\{O\} = |O| - E$ will be uniformly zero. However, in all other (non-uniform) circumstances, $|O^{+}|$ realization is always lucky (even when negative!)

* * *

There is what on casual view looks to be something of an anomaly, namely that it is unlucky for us to be in a condition of affairs where a lucky occurrence is required, and lucky to be in a circumstance where a need lucky occurrences are dispensable. But second thought will—or should indicate that this is just as it ought to be.

To see that one is unlucky if realizing one's goal requires a good deal of luck. Consider:

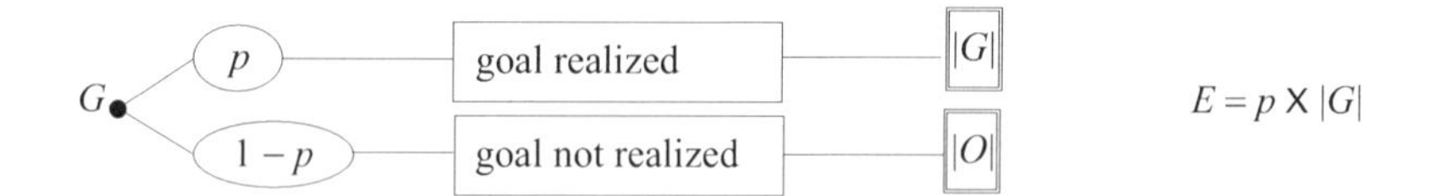

Here we have

$$\lambda\{G\} = Y - E = |G| - p \times |G| = |G| \times (1 - p).$$

So if reaching the goal required much luck, then $|G| \times (1 - p)$ must be sizable. So $1 - p$ will have to be fairly large (i.e., close to 1), which renders the probability of goal achievement unlikely. To be in such a situation having to rely on luck and make demands for its service is a rather unhappy state of things.

* * *

Throughout, in chancy situations, the expectation will always fall between the worst-possible and the best-possible outcomes:

$$|O^-| \leq E \leq |O^+|$$

Accordingly

$$|O^-| - E \leq E - E \leq |O^+| - E$$

and

$$\lambda\{O^-\} \leq O \leq \lambda\{O^+\}$$

As long as there are outcomes of different yields, there will be lucky and unlucky one.

Consider a chancy situation whose structure is

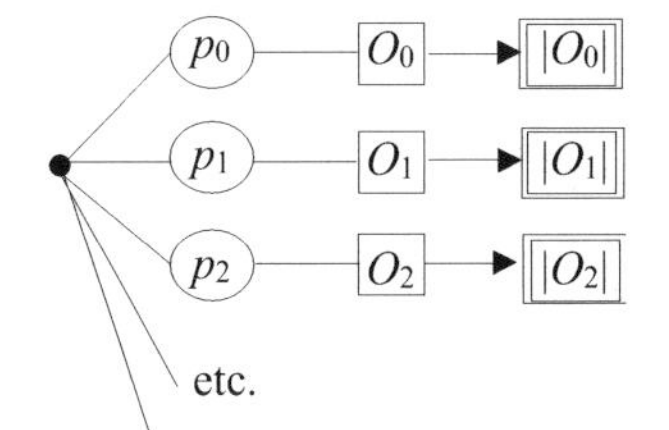

Let us inquire into the luck of not-O_0. This would, of course, be available via $\lambda\{\text{not} - O_0\} = |\text{not} - O_0| - E$. Here E is no problem. But what is $|\text{not} - O_0|$?

Here we can embark on what is effectively a reduction of the $|O_0|$ box.

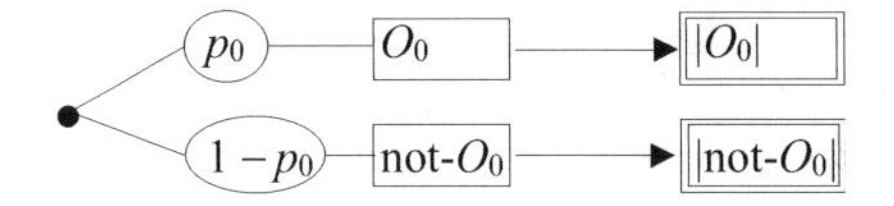

And since $E = p_0$ and $E - p_0 \times |O_0| + (1 - p_0) \times |\text{not-}O_0|$ we have

Display 2.1

A LUCK COMPARISON

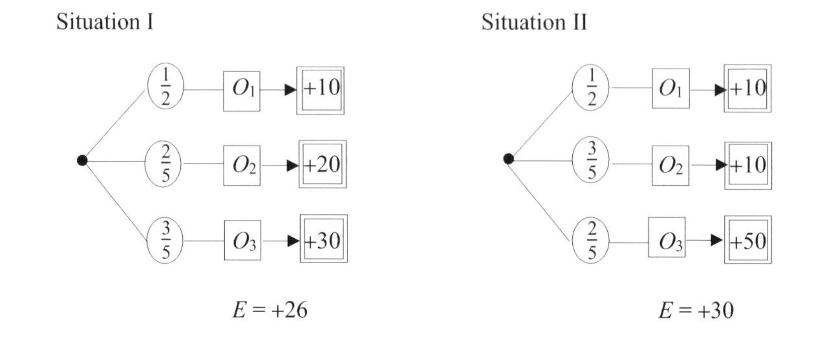

$$|\text{not-}O_0| = \frac{\mathrm{E} - \mathrm{p}_0 \times |\mathrm{O}_0|}{1 - p_0}$$

Accordingly, since $\lambda\{\text{not-}O_0\} = |\text{not-}O_0| - E$ we have:

$$\begin{aligned}
\lambda\{\text{not-}O_0\} &= \frac{\mathrm{E} - p_0 \times |O_0|}{1 - p_0} - E \\
&= \frac{\mathrm{E} - p_0 \times |O_0| - E + E \times p_0}{1 - p_0} \\
&= \frac{-p_0(|O_0| - E)}{1 - p_0} \\
&= \frac{p_0}{p_0 - 1} \times \{O_o\}
\end{aligned}$$

* * *

Luck is a fundamentally contextual business; it hinges not on outcome value but on alternative possibilities. Thus consider the situation depicted in Display 2.1, when yields of these three outcomes are exactly the same.

Note that $\lambda = Y - E$ means that in Situation I the luck of outcome O_1 is -16, while in Situation II the luck of outcome O_1 is a significantly smaller -20. For while the same outcome with the same probability is at issue, the difference in context becomes critical.

* * *

Consider the following chancy situation:

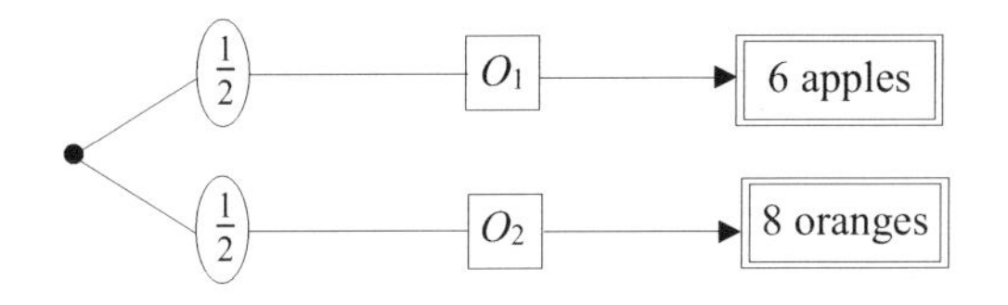

With $|O_1| = 3$ apples and $|O_2| = 4$ oranges we cannot proceed to assess luck unless these outcome-results themselves are specified commensurably in a common unit of value. Only if told something along the lines that 6 apples cost $2 while 8 oranges comes $4 can we say that it is luckier to win the oranges. (For only now does $\lambda = Y - E$ enable us to obtain $\lambda\{O_1\} = -1$ and $\lambda\{O_2\} = +1$ so that in winning the oranges you are $2 luckier than with the apples—exactly as one would then expect).

* * *

Clearly, then, the quantitative assessment of luck has certain prerequisites and so becomes practicable only if certain conditions and presuppositions are met. Specifically we must be able to:

- understand the situation at issue sufficiently to determine the range of possible outcomes exhaustively.
- comprehend these outcomes in sufficient detail so as to be able (1) to evaluate the direction and extent of their bearing on the interests at issue, and (2) to assess or estimate the likelihood (or probability) of their realization.

Only where these requirements are satisfied will a quantitative approach to the assessment of luck become possible.

* * *

How much luck is afforded overall by a chancy situation? With $\lambda\{O_i\} = |O_i| - E$ we have:

$$\sum_i \lambda\{O_i\} = \sum_i (|O_i| - E) = \sum_i |O_i| - i \times E$$

The overall total of the luck afforded by a chancy situation is thus the difference between the totality of its outcome yields minus the appropriate multiple of its expectation.

And this in turn means that:

$$\frac{1}{i} \sum \lambda\{O_i\} = \frac{1}{i} \sum |O_i| - E$$

The average luck of the outcomes a chancy situation is its average yield minus the expectation. (In fair gamble where $E = 0$ this becomes an identity.

* * *

Three sorts of anomalous chancy situations have characteristic implications:

(1) *When all outcomes are equiprobable, then the expectation will be* $E = p \times \sum_{i}^{n} |O_i|$, and thus a fixed constant for the entire situation. This enjoins $\lambda\{O_i\} = |O_i| - [\text{const}]$. The luck of an outcome will thus be quasi-proportional to its yield.[3]
(2) *When one outcome probability is 1, so that all the rest have probability* 0, then $E = |O|$ for that uniquely possible outcome. So then always $\lambda\{O\} = 0$. Inevitable outcomes have zero luck.
(3) *When the expectation of the chancy situation is zero*, ($E = 0$), then the luck of an outcome is simply that outcome's yield: $\lambda\{O_i\} = |O_i|$, uniformly for all i.

Luck thus behaves in a characteristic way in each one of these anomalous situations.

* * *

When you play Smith at billiards, you win x percent of the time. How lucky will you be to win the next time? With your win probability at x percent, the win expectation will be $\frac{x}{100}$ and your win-luck will thus be $1 - \frac{x}{100}$; i.e., will exactly equal your loss probability.

But suppose that you win $y\%$ of your games by skill (and so $(100 - y)$ percent of them by luck). How probable is it now that you will win your next game by luck? Well—you win x percent of the games and y percent of these by luck, so you'll likely win $x \times y$ percent of them by luck, and if this holds generally then your chance of winning the next game by luck is $x \times y$ percent. But how lucky is it for you to win that next game? With (win - luck) = (loss probability) the answer is: $\lambda\{\text{win}\} = 1 - \frac{(1-x)}{100} = \frac{x}{100}$.

* * *

Two evenly matched players square off against each other. Assume that one wins a match by skill so that we have skill – 1, luck – 0. But his luck in realizing this results is $\lambda = Y - E = 1 - \frac{1}{2} = \frac{1}{2}$. So clearly two very different things are at issue, viz.

performance or process luck

in the first case, and

outcome or product luck

in the second. The former answers the question: How was the result achieved: by skill or cheating or luck or . . . ? The latter answers the questions to what extent does the result provide a benefit?

To get a firmer grip on the distinction consider A playing a four-game match against an evenly able B, with the following results (here: X = loss, S = win by skill, and L = win by luck):

		A's Result
Case 1:	$L\ S\ X\ L$	11
Case 2:	$S\ S\ L\ X$	7

The result scoring is to exponential: 1 point for a game 1 victory, 2 for game 2, 4 for game 3, and 8 for game 4 (nothing for a loss). On this basis, we have A's result as indicated. And its expectation is $E = 7\frac{1}{2}$. Question: Which case is the luckier for A? Note that:

(1) As to performance luck Case 2 prevails. Here A (by hypothesis) wins twice times by luck rather than overly once as in Case 1.
(2) But as to product luck we have $\lambda = Y - E$ is $+3\frac{1}{2}$ in Case 1 and $-\frac{1}{2}$ in Case 2. So Case 1 now prevails.

All depending on the circumstances, winning by luck need not be all that lucky.

The important point for present purposes is that decidedly different matters are at issue with outcome luck and luck in performance, and that our present metric account of luck is oriented in the former direction.

* * *

In chancy situations, the *luck-differential* of an outcome O is defined as:

$$\mathcal{L}\{O\} = \lambda\{O\} - \lambda\{\text{not} - O\}$$

Given that the λ-luck of an outcome is the difference between its yield and the situatinal expectation ($\lambda = Y - E$), we obtain:

$$\mathcal{L}\{O\} = |O| - |\text{not} - O|$$

But from $E = p \times |O| + (1 - p) \times |\text{not - } O|$, where $p = pr\{O\}$, we obtain

$$|\text{not} - O| = \frac{1}{1 - p} \times [E - p \times |O|]$$

Accordingly:

$$\begin{aligned}\mathcal{L}\{O\} &= |O| - \frac{1}{1 - p} \times [E - (p \times |O|)] \\ &= \frac{(1 - p) \times |O| - [E - (p \times |O|)]}{1 - p} \\ &= \frac{|O| - E}{1 - p} = \frac{1}{1 - p} \times \lambda\{O\}\end{aligned}$$

In sum, the luck differential of an outcome is simply a probability-qualified variant of its luck. The concept thus contributes little that is substantially new.

Recall, moreover, that a few pages ago we saw that $\lambda\{\text{not-}O\} = \frac{p}{p-1} \times \lambda\{O\}$. So now with $\mathcal{L}\{O\} = \frac{1}{1-p} \times \lambda\{O\}$ we obtain:

$$\frac{p}{1-p} \times \lambda\{\text{not-}O\} = (1-p) \times \mathcal{L}\{O\}.$$

Accordingly, we also have it that $\mathcal{L}\{O\} = \frac{-1}{p} \times \lambda\{\text{not-}O\}$.

Question Appendix

- *Question 1*: Under what conditions are all of the possible outcomes of a chancy situations equally lucky?
  Answer: If $\lambda\{O_i\} = c$ uniformly for all i, then uniformly $|O_i| - E = c$ and thereby $|O_i| = E + c =$ constant. Thus only when all outcome-yields are identical will all outcomes be equally lucky. Probabilities do not enter into it then.
- *Question 2*: What is the sum total of outcome luck afforded by a chancy outcome situation?
  Answer: $\sum_i^n (|O_i| - E) = \sum_i |O_i| - n \times E$. So when we have is a fair situation ($E = 0$), then the luck total will equal the yield total. Otherwise things get complicated.
- *Question 3*: Can it happen that (a) Every outcome of a chancy situation is lucky, or (b) that no outcome is lucky.
  Answer: (a) Clearly not. It cannot be that $\lambda\{O_i\} = |O_i| - E$ is always > 0. For this would require that all $|O_i|$ are $> E$ which is impossible given E's definition. (b) However, when all outcomes have equal yield then none will be lucky.
- *Question 4*: Can it happen that (a) Every outcome of a chancy situation is unlucky, or (b) that no outcome is unlucky?
  Answer: (a) clearly not. It cannot be that $\lambda\{O_i\} = |O_i| - E$ is always < 0. For this would require that all $|O_i|$ are $< E$ which is impossible given E's definition. (b) However, when all outcomes have equal yield then none will be lucky.
- *Question 5*: Under what condition will an outcome-yield in a chancy situation equal the expectation of that situation itself?
  Answer: Consider a situation of the following format:

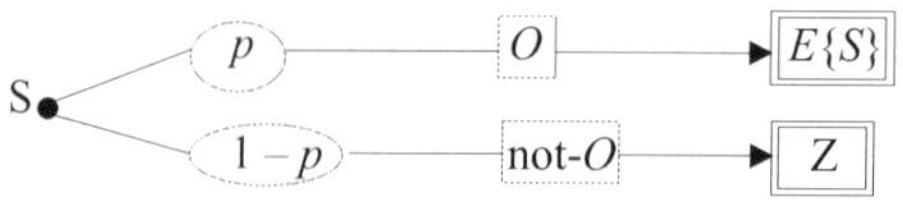

S → p → O → $E\{S\}$
S → $1-p$ → not-O → Z

$$E\{S\} = p \times E\{S\} + (1-p) \times Z$$

From the specification of $E\{S\}$ it follows that $E\{S\} = p \times E\{S\} + (1-p) \times Z$, and so: $E\{S\} = Z$. The value of p become irrelevant, and $Z = E\{S\}$. The answer to our question in thus: as long as all the outcomes of the situation are identical.

Notes

1. Yet one certainly should not identify luck and chance, because most chance developments in nature exert no effect on anyone for good or for ill. Only with matters of chance where when someone's interests are at stake does luck come into play.
2. The probabilities at issue in these expected-value assessments should, of course, be the prior probabilities obtaining in advance of the fact. The ex-post-facto probability that a head-yielding toss has yielded a tail is a here-irrelevant zero.
3. Recall that $f(x)$ are $g(x)$ are quasi-proportional when there are constants c and k such that $f(x) = c \times g(x) + k$.

Chapter 3
Variant Approaches to Luck

The standard approach to assessing luck will proceed via the amount-measure of the Basic Luck Equation $\lambda = Y - E$. However, some variant perspectives bring other salient features of luck to light.

Consider the situation of a die-toss where you receive \$5 for every point by which the outcome exceeds 1. You roll the die and get a two, benefitting you by \$5. But that second-worse outcome certainly does not make you particularly lucky. Seeing that you could just as likely have scored \$25 it will not have you "thanking your lucky star." How, then, does luck work.

Evaluating the outcomes of stochastic situations is a challenging business because different sort of merit address different albeit related aspects of the matter. Clearly we must acknowledge the contrast between absolute and comparative luck. A quality like "80% of the way from worst possible outcome to the best possible" certainly measures something and is unquestionably indicative of something relating to merit. But it takes no account of chance and likelihood despite its bearing on good fortune is certainly no adequate measure of luck.

* * *

It may seem tempting on first thought to consider assessing the luck of an outcome via greatest possible yield. But a simple illustration suffices to show that this will not quite do. For consider the following distribution of outcome values alone a number axis:

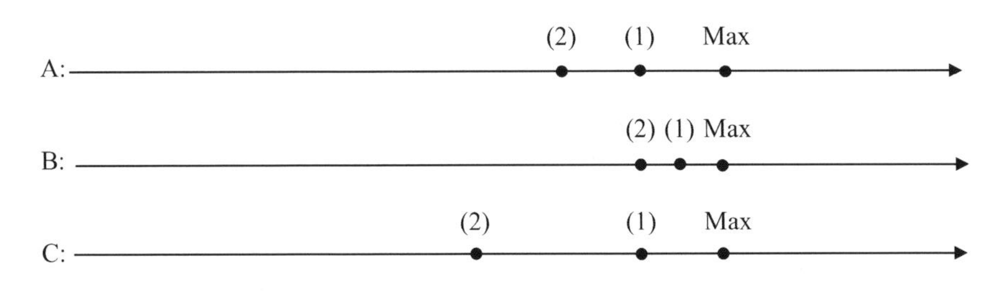

N. Rescher, *Luck Theory*, Logic, Argumentation & Reasoning 20,
https://doi.org/10.1007/978-3-030-63780-4_3

In comparing A and B we note that outcome $B2$ is closer to their common Maximum than $A2$ is, but yet one would not be included to see $B2$ as well that lucky because it is the worst outcome in its group. And in comparing $A2$ with B2, we see that while $C2$ is inferior to both, yet we would deem $A2$ as unlucky because it is the worst outcome in its group and yet see $C2$ as rather lucky because it averts the misfortune of at $C2$.

* * *

A promising way of assessing the luck of an outcome is via its relative placement between the worst-possible outcome $|O^-|$ and the best possible $|O^+|$. This can, of course, be done in two ways, namely via proximity to $|O^+|$ or via distance from $|O^-|$. The former approach leads to disappointment luck:

$$\lambda^+\{O\} = \frac{|O^+| - |O|}{|O^+| - |O^-|} = \frac{|O^+| - |O|}{\Delta}$$

(Here $|O^+| - |O|$ reflects outcome shortfall from the maximum—pf course the less the better. And the latter approach leads to the measure of loss avoidance

$$\lambda^-\{O\} = \frac{|O| - |O^-|}{|O^+| - |O^-|} = \frac{|O| - |O^-|}{\Delta}$$

(Here that $|O| - |O^-|$ reflects loss-avoidance—of course the more that better. It ranges from 0 at $O = O^-$ to 1 at $O = |O^+|$. What λ measures that actual *amount* of luck λ^- only assesses the comparative *degree* of luck.

* * *

All of these variant luck measures are quasi-proportional to the basic yield luck λ. (Recall that the formulas f and g are quasi-proportional when there are constructs c and k such that $f(x) = c \vee g(x) + k$. For example consider:

$$\lambda^-\{O\} = \frac{|O| - |O^-|}{\Delta} = \frac{|O| - |O^-| + E - E-}{\Delta}$$
$$= \frac{\lambda|O| - \lambda|O^-|}{\Delta} = \frac{1}{\Delta} \times \lambda\{O\} + \frac{\lambda^-\{O\}}{-\Delta}$$

Here both $\frac{1}{\Delta}$ and $\frac{\lambda^-\{O\}}{-\Delta}$ are fixed constants for the chancy situation at issue.

And a similar story holds for the other luck indices as well. All are quasi-proportional to λ-luck and this introduces nothing new at the level of fundamentals.

Accordingly, an instructive example of luck's modus operandi is conveyed in Display 3.1, where the λ column addresses the quantitative *amount* and orientation of luck and the λ^- column its comparative *extent.* (Observe the discrepancy around the 15 year mark owing to the lesser precision of λ^-.)

Display 3.1

MATRIMONIAL LUCK

(Based on the Percentage of Living Ruritanian Couples Still Married after *Y* Years in 1920)

Married Years (Y)	% Yet Married	Probability of Duration-Group Membership (p)	Expectation	λ	λ^-
5	85%	.30		–9	.17
10	70%	.25		–4	.33
15	15%	.18	14	+1	.5
25	40%	.14		+16	.8
35	35%	.12		+22	1

That $\lambda\{O\}$ and $\lambda^-\{O\}$ are quasi-proportional means that within any chancy situation both will always yield the same rank order for outcomes, and answer the question "Is O_1 luckier than O_2?" in the same way. Numerically they will differ, but rank-wise they will agree. For λ^- reflects position within the outcome spectrum Δ relative to the merely quantitative (arithmetical) mid-point between $|O^-|$ and $|O^+|$, λ, by contrast, reflects position relative to the (probabilistic) center of gravity.

* * *

The prime considerations in assessing outcome luck is how the possible outcomes are distributed overall in point of value. For clarity here consider Display 3.2. The formative idea for luck is that one is lucky in achieving an outcome (result) that is more than middling in merit. But what is to qualify as the determinative middle? There are three possibilities for locating an outcome between the worst and the best, between $|O^-|$ and $|O^+|$.

(1) the *arithmetical middle* between these extremes, namely $\frac{1}{2}(|O^+| - |O^-|)$

(2) the *numerical average*(i.e., arithmetical mean) of the outcome yields: $\frac{1}{n}\sum_{i}^{n} |O_i|$

Display 3.2

ILLUSTRATIVE MAP OF THE OUTCOME RANGE

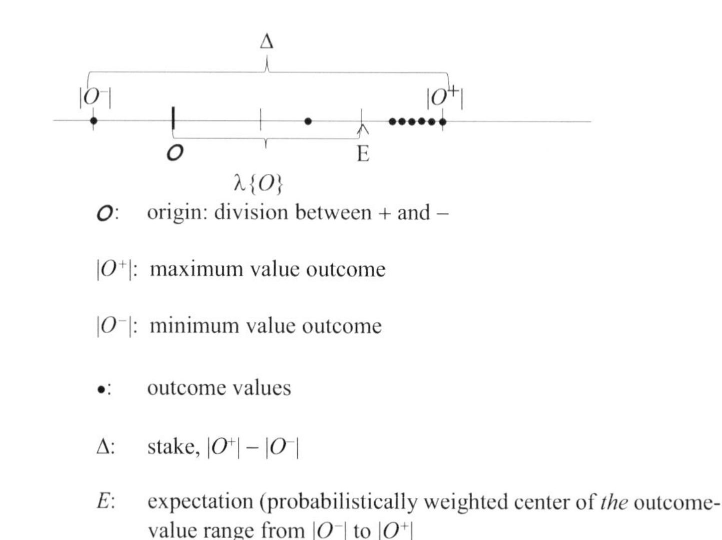

(3) the probabilistically weighted average of the $|O_i|$, namely the *expectation*:

$$E = \sum_{i}^{n} (p_i \times |O_i|)$$

Since (3) is clearly a more sophisticated construal because it takes the probabilistic aspect of the distribution into account. If only a comparative assessment of outcome luck is needed, a crude standard such as λ^{-} would suffice, but if actual amounts are warranted then one will need to use λ. In the end, we thus have two different evaluations for luck: either via its (*sortal*) *amount*—as measured by λ—or via its (dimensionlessly *comparative*) *extent* as per the index λ^{-}.

* * *

Over and above standard mode of luck which is basically *gain luck* (λ) and hinges on comparing the actual outcome $|O|$ with the expectation—as per the Basic Luck Equation $\lambda\{O\} = |O| - E$—there is also the rather different conception of what might be called *loss-avoidance luck*. This is a matter of comparing the actual outcome $|O|$ with what would happen "if worst came to worst," and would accordingly be measured by $|O| - |O^{-}|$. This quantity will also amount to $(|O| - E) - (|O^{-}| - E) = \lambda\{O\} - \lambda\{O^{-}\}$. With $\lambda\{O^{-}\}$ a situational fixity this means that loss luck is quasi-proportional to gain luck. Nothing that is fundamentally new form a conceptual standpoint is at issue.

Introduction of this idea does, however, for a disambiguation of luck talk. Thus someone engaged in a chancy situation who gets a result of -10 where the expectation was zero, is unlucky in the standard sense of gain-luck, but also very lucky in the manner of loss-avoidance if an outcome of -100 had figured among the possibilities. While only seldom acknowledged, the distinction between gain-luck and loss-luck demands heed through the range of luck-attribution. However, the former concept is so much more familiar that it has become dominant.

* * *

Moreover, there is yet another approach to luck assessment. It is instructively approached via the question: How lucky is it to be lucky?

Interestingly, it appears that luck is not reflexive—that it is not always λ-lucky to be λ-lucky. Thus consider a situation with twelve equiprobable outcomes whose yields are distributed as follows.

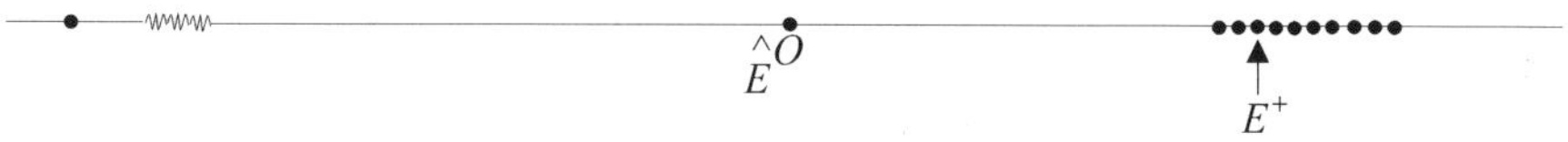

$\hat{O}$
E E^+

By the standard of $|O| > E$ that central outcome O is certainly lucky: its yield is above expectation. However, its standing in the range of possible outcomes is dismal, since all but one alternative outcome is distinctly superior to it. Its recipient has every right to be disappointed.

However, rather than invalidating the present approach to luck this example highlights the complexity of the idea in having different aspects that address different issues. For even a lucky outcome can be seen as comparatively unlucky when there are many superior alternatives. Consider in this light the idea of the *second-order luck* of an outcome via its standing within the range of otherwise lucky outcomes. In the overall range of lucky outcomes that lucky outcome O fares the very worst—and thereby is *second-order unlucky*.

In this light the second-order luck standing of an outcome O can be assessed at:

$$\lambda^{\#}\{O\} = |O| - E^{+}$$

where E^+ is the positive subsector of E. (Note that in the above example that "lucky" outcome O is decidedly unlucky.)

First-order λ-luck and this second-order $\lambda^{\#}$-luck will go their own separate ways: one can be λ-lucky (with $\lambda\{O\} > 0$) and nevertheless $\lambda^{\#}$ unlucky (with $\lambda^{\#}\{O\}$ comparatively small) seeing that different qualifying standards for being "lucky" are in play. Of course what one would ideally want in a chancy situation is an outcome that is lucky in both orders.

Such first and second-order luck are both modes of luck, one might well ask: How is it that these two standards can get out of synch in point of having luck? The answer is that they qualify being lucky by two very different criterions, viz., "being positive" for λ, and being comparatively large for $\lambda^{\#}$).

3.1 Summary

To summarize.

The proceeding have introduced various modes of outcome luck along with their corresponding ways of assessment. Two modes are absolute:

I. *Yield luck*. The excess or shortfall over expectation (E):

$$\lambda\{O\} = |O| - E$$

Potentially λ ranges from $-\infty$ to $+\infty$. And outcome qualifies as lucky when $\lambda\{O\} > 0$.

II. *Second-order luck*: The standing of a lucky outcome within the overall range of lucky outcomes:

$$\lambda^{\#}\{O\} = |O| - E+$$

Here E^* is the positive subexpectation of E. Potentially $\lambda^{\#}$ ranges from 0 to $|O^{+}|$. And outcome qualifies as second-order lucky when $\lambda^{\#} > E^{*}$.

There are also two comparative standards. Both look to the comparative place of the outcome in the range from $|-O|$ to $|+O|$.

III. *Loss-averting luck*: Reflecting the distance of the outcome from the worst possibility the extent of its opportunity:

$$\lambda^{-}\{O\} = \frac{|O| - |O^{-}|}{|O^{+}| - |O^{-}|}$$

Potentially it ranges from 1 to 0—the smaller the better, An outcome qualifies as lucky when $\lambda^{+} < \frac{1}{2}$.

IV. *Disappointment luck*, reflecting the outcome's shortfall from the maximum:

$$\lambda^{+}\{O\} = \frac{|O^{+}| - |O|}{|O^{+}| - |O^{-}|} = \frac{|O^{+}| - |O|}{\Delta}$$

Potentially it ranges from 1 to 0. An outcome qualifies as lucky when $\lambda^{-} < \frac{1}{2}$.

Chapter 4
Some Illustrative Examples

Luck has a double aspect (1) as facilitating an outcome's realization, and (2) as consequent on the outcome realized. Thus as to (1), the burglar was lucky that the watchman was asleep thus greatly easing his work. And as to (2) he was lucky that the house he targeted unbeknownst to him happened to contain a large collection of gold coins. Thus there can be luck in production or luck in product. And the two are connected. When you are lucky in obtaining what you get, then it is lucky for you *that* you got it.

* * *

The Basic Luck Equation $\lambda\{O\} = |O| - E$, has various important consequences:

1. When outcome equals expectation and $|O| = E$, then we have $\lambda = 0$. When the outcome is exactly what is to be expected, there is no luck (be that expectation positive or negative).
2. In a given chancy situation luck increases with increasing outcome-yield: $\lambda\{O_i\} = |O_i| +$ (constant). Accordingly, an outcome of maximal yield $|O^+|$ is ipso facto bound to be the luckiest, regardless of the likelihoods involved. And analogously with $|O^-|$.
3. With $E = 0$, $\lambda\{O\} = |O|$. In fair gambles—those with $E = 0$—the luck of an outcome is simply that outcome's yield itself.
4. With $|O| = 0$ we have $\lambda = -E$. When an outcome has zero yield, the luck of its realization is simply the negative of the situational expectation.

* * *

Let it be that someone is so situated that they win a dollar for each point in a doe-toss, so that their expectation is:

N. Rescher, *Luck Theory*, Logic, Argumentation & Reasoning 20,
https://doi.org/10.1007/978-3-030-63780-4_4

$$\$1 \times \frac{1}{6} + \$2 \times \frac{1}{6} + \ldots + \$6 \times \frac{1}{6} = \frac{\$21}{6} = \$3.5$$

Then if the result of the toss is 1 the agent is unlucky to the tune of \$1 – \$3.50 = – \$2.50. And if the result is 6 the agent is lucky to the tune of \$6 – \$3.50 = \$2.50.

* * *

Again consider the following chancy-outcome situation:

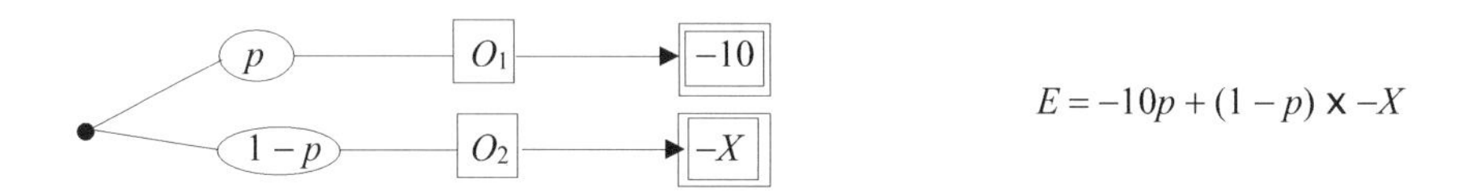

Of course the protagonist is not going to be particularly pleased with O_1's loss of 10; no-one welcomes a loss. But all the same, one would certainly deem oneself lucky in averting the loss of any sizable $-X$. But just how lucky would this be?

By $\lambda = Y - E$ we have:

$$\lambda\{O_1\} = -10 - E$$

And

$$\lambda\{O_2\} = -X - E$$

With $X > 10$ outcome O_1 is going to be luckier than O_2, quite independently of the value of p. Averting a larger rather than lesser loss is always good luck.

* * *

One is not necessarily lucky in achieving a positive outcome. In a coin-toss where Heads is to yield a \$100 award and Tails one of \$1,000, we would deem the \$100 winner unlucky in losing out on the bigger prize. In falling far short of the expectation (of \$550) is not just a matter of the outcome value itself but of its alternatives. If you win \$2 when you might equally well have won \$10 you are comparatively unlucky; if the best- available alternative had yielded only \$1, you are comparatively lucky. The luck of a given outcome is not determined by this result but is also—and critically—a matter of the alternatives. It is lucky for a disaster to be avoided, even if at a price.

One is not necessarily unlucky in taking a loss. For the luck equation has it that one cannot say that someone is unlucky in realizing an *expected* loss. Is this a problem? Suppose John knows from the outset that the race-horse "Danny Boy" has been ailing and cannot win the race. He bets on him anyway. And, of course, loses. Is he unlucky in doing so? Tom knows that Jack is a cheat and a scandal. He lends him the money anyway. And finds that it isn't repaid. Is he not unlucky in suffering this business loss? Foolish no doubt, unfortunate perhaps. But certainly not unlucky. The reality of it is that luck comes into it only where chance is at work and is not on the scene when the outcome is a forgone conclusion.

With luck operative not just within but also as between situations this second mode of evaluation becomes a requisite.

* * *

The tabulation of Display 4.1 presents the luck picture of the three hypothetical stochastic outcome situations, consisting of three main components: an inventory of the total spectrum of possible outcomes, a tabulation of their respective probabilities and yield, and a specification of the amount of λ-luck afforded by the various outcomes.

Display 4.1

SOME LUCK ILLUSTRATIONS
(Three Variant Cases of the Results of a Coin Toss)

Outcome	Probability	Yield/Expectation/Luck ($\lvert O_i \rvert/E/\lambda_i$)		
O_i	p_i	Case 1	Case 2	Case 3
1	$\frac{1}{6}$	6/24/−18	6/42/−36	30/45/−15
2	$\frac{1}{6}$	6/24/−18	6/42/−36	30/45/−15
3	$\frac{1}{6}$	6/24/−18	60/42/+18	30/45/−15
4	$\frac{1}{6}$	6/24/−18	60/42/+18	60/45/−15
5	$\frac{1}{6}$	60/24/+36	60/42/+18	60/45/+15
6	$\frac{1}{6}$	60/24/+36	60/42/+18	60/45/+15

Display 4.1 further illustrates some key aspects of luck. Specifically: That there are good-luck-predominant situations and bad-luck-predominant ones, and that in equal-probability-outcome situations the sum total of luck is always zero, so that here the good luck of one outcome can have only be realized here at the expense of the bad of another. Moreover, we can see that (as outcome 3 shows) situations where the same outcome is yielded with the same probabilities, differences of stochastic context can endow that outcome with very different modalities of luck.

* * *

Consider a chancy situation regarding goal realization:

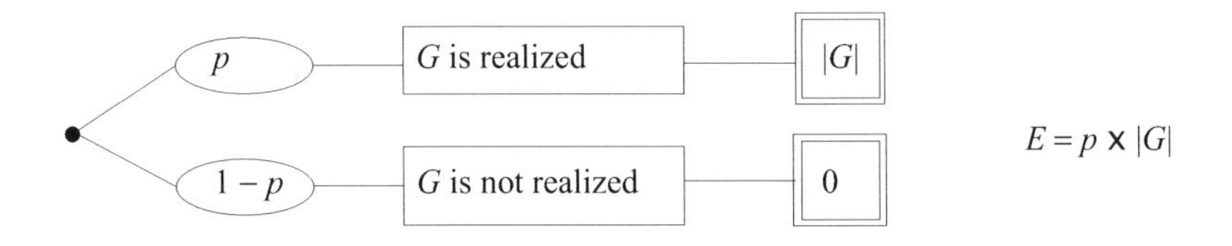

If reaching your goal G requires a good deal of luck, then $\lambda\{G\} = |G| - E$ is substantial. But so $\lambda\{G\} = |G| - p \times |G| = |G| \times (1 - p)$ cannot be substantial when p is close to 1. So it is not a happy circumstance for substantial luck to be required for reaching a goal. To have been lucky is always a good thing, but to require luck for goal realization is not. Luck wears a different mean relationship and prospectively.

* * *

It is instructive to examine luck in situations of conflict and competition. Thus compare two contestants engaged in a competitive public match, one (X) with a better performance record and more "drawing power" than the other (Y). They arrange of payment of 16% of the "gate," to be divided as follows:

The Victor	X gets	Y gets
X	10%	6%
Y	8%	8%

X is the likely victor with probability 60% we have it with respect to expectation that (in percentage terms):

$$E_x = .6 \times 10 + .4 \times 6 = 8.4$$

$$E_y = .4 \times 8 + .6 \times 8 = 8.0$$

In consequence we have:

$$\lambda_x\{\text{win}\} = 10 - 8.4 = +1.6$$

$$\lambda_x\{\text{lose}\} = 6 - 8.4 = -2.4$$

$$\lambda_y\{\text{win}\} = 8 - 8.0 = 0$$

$$\lambda_x\{\text{lose}\} = 8 - 8.0 = 0$$

Owing to the difference in expectation, X's unluck in losing far exceeds X's.

* * *

Again, let it be that the weather forecast indicates a 30% chance of rain. Smith wears his new \$120 hat, which will be ruined if rained on. How lucky/unlucky will he be according on how things work out?

As regards expectation we have:

$$E = (\text{Rain yield}) \times \frac{30}{100} + (\text{Non-rain yield}) \times \frac{70}{100}$$

Since Smith's rain yield is a \$120 loss and the no-rain yield is zero, his expectation will be $E = -\$36$. So on this basis Smith's luck will be:

$$\lambda\{\text{rain}\} = -120 - (-36) = -\$84$$

and

$$\lambda\{\text{no rain}\} = 0 - (-36) = +\$36$$

So if it rains, Smith is quite unlucky, if it does not he is somewhat lucky. And the difference is exactly the value of that $120 hat. The luck equation clearly produces the intuitively right result. Note, however, that the rain luck is *not* equal and opposite of the no-rain luck.

* * *

Good luck can come in tragic—even catastrophic—guise. Suppose that a plague comes to the land, devastating its population. Virtually every family has its victims, as per Display 4.2. Since an outcome qualifies as lucky when $\lambda > 0$ it emerges that (in the tragic circumstances) it is lucky for a family to lose even as much as a quarter of its members.

Display 4.2

A HYPOTHETICAL PLAGUE IN 1580 RURITANIA

% loss in Family Members (Y)	% of Families with such loss	Probability of a random family's loss-group membership	E (in % of Y)	$\lambda = Y - E$
0	15	.15		+36
−10	10	.1		+26
−25	30	.3	−36	+11
−50	30	.3		−14
−75	10	.1		−39
−100	5	.05		−64

---

* * *

Three distributions of possible outcomes can prevail in a chancy situation, respectively those that are

I. evenly balanced between good-luck outcomes and bad-luck outcomes.
II. predominated by good-luck outcomes
III. predominated by bad-luck outcomes

An illustration is provided in the luck profile of Display 4.3 with its two very different luck distributions based on a die toss. Note that in each case having lucky/unlucky outcomes will be lucky/unlucky respectively.

DISPLAY 4.3

THE INFLUENCE OF OUTLIERS

Outcome	Probability	Yield Case #1	Yield Case#2	Expectation #1	Expectation #2	Luck ($\lambda = Y - E$) #1	Luck ($\lambda = Y - E$) #2
1	$\frac{1}{6}$	+19	−19	+4	−4	+10	−15
2	$\frac{1}{6}$	+1	−1			−3	+3
3	$\frac{1}{6}$	+1	−1			−3	+3
4	$\frac{1}{6}$	+1	−1			−3	+3
5	$\frac{1}{6}$	+1	−1			−3	+3
6	$\frac{1}{6}$	+1	−1			−3	+3

Note too that in Case #1 the likely-hood of an unlucky outcome predominates (notwithstanding the pervasiveness of positivity). And in case #2 the opposite is the case. In case #1 most outcomes represent a lost blessing; in case #2 most represent an averted misfortune. As the example shows, a situation where the prospect of good luck predominates is not always to be welcomed. Luck and benefit are different issues.

* * *

The occurrence of a *near miss* is a common sort to luck. Here the protagonist is caught up in circumstances where one would ordinarily hope for a certain positive (or negative) outcome, but fails by a narrow margin to realize (or avoid) it. For example, you might need to win 100 points to qualify for some benefit, but realize only 99. The luck situation obtaining with near misses illustrated by Display 4.4. In this illustration the luck associated with the three outcomes via $\lambda = Y - E$ will be:

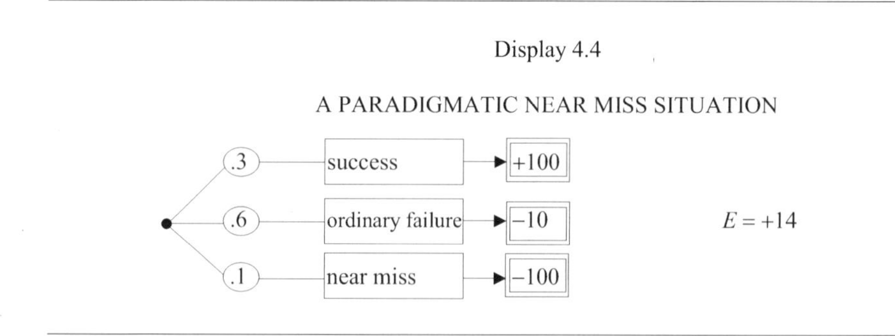

$$\lambda\{\text{success}\} = +86$$

$$\lambda\{\text{ordinary failure}\} = -24$$

$$\lambda\{\text{near miss}\} = -114$$

The situation assumes that in the circumstances success is in prospect, through failure likely, and that a near miss, though unlikely, is particularly frustrating. Such a near miss is to be seen as particularly unlucky (unluckier even than success is lucky.)

However, if that near miss were not appraised quite so negatively, say by scoring only −60, then we would have $E = +18$ and those three luck scores would be:

$$\lambda\{\text{success}\} = +82$$

$$\lambda\{\text{normal failure}\} = -42$$

$$\lambda\{\text{near miss}\} = -76$$

The luck difference between an ordinary miss and a near miss would now decline from 90 to 32.

* * *

The theme of *lucky escapes* poses interesting issues. Be it from prisons, from dictatorships, or from enemy prisoner stockades, escape accounts exert an ongoing fascination. The expression is almost a redundancy seeing that in the circumstances successful escape is the very paradigm of good luck, in realizing a huge yield—years of lifetime or sometimes the majority of it. With these narrow escapes the possibility of success is generally contrived to render escape difficult if not impossible, and so that the expectation of success will be very small. In consequence, luck as determined by $\lambda = Y - E$ is large. The escape itself may be a ruse of elaborate planning and painstaking management that "leaves nothing to luck," but its success is lucky all the same.

With a *lucky escape* one realizes one's goal $|O|$ in circumstances where $|\text{not-}O|$ is far more negative than $|O|$ (hence "escape") and where p is rather small (hence "narrow"). With a *missed opportunity*, by contrast, one realizes the outcome O in circumstance where not-O is far more positive and where is rather small (hence "opportunity") p is smallish but the comparative size of $|O|$ and $|\text{not-}O|$ substantially reversed.

For the sake of illustration, consider Display 4.5.

Comparing the expectations here shows how—as we would expect—lucky escapes are indeed quite lucky, while lost opportunities are quite unlucky.

Display 4.5

TWO CONTRASTING LUCK ILLUSTRATIONS

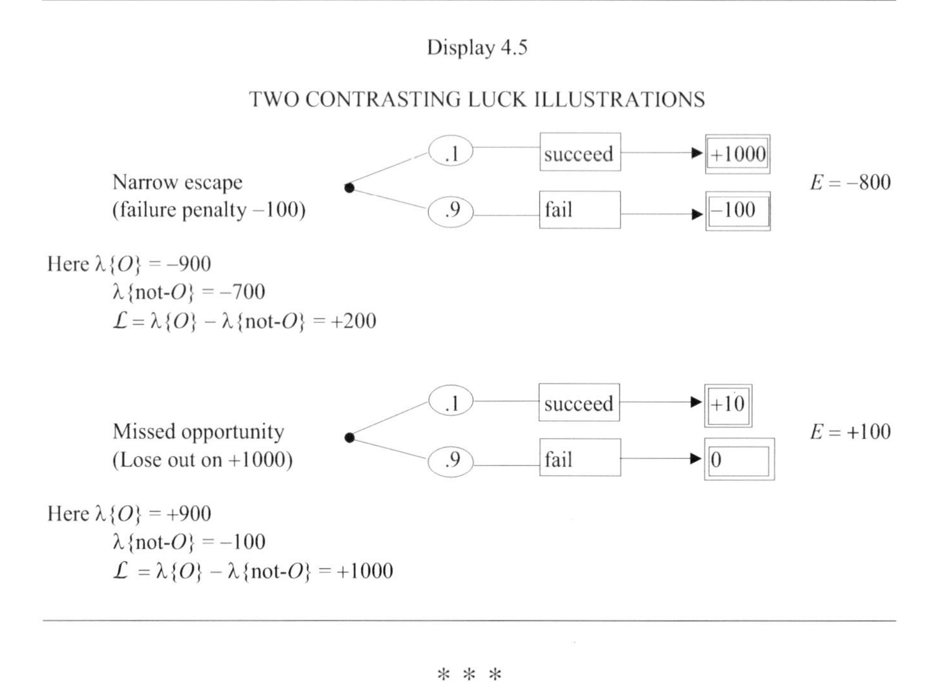

* * *

Again suppose that you fall ill, are treated by a doctor, and recover rapidly. In the end you are lucky. But to what extent? Consider Display 4.6. There are two possibilities here:

(1) The doctor who treated you is a specialist who succeeds in 95% of cases of your sort. So now with V as the value of your life, the luck formula $\lambda = Y - E$ provides for:

$$\lambda_1 = V - .95V = .05V$$

(2) The doctor who treats you is a non-specialist who accordingly succeeds in 80% of cases of your sort. So now

$$\lambda_2 = V - .8V = .2V$$

Clearly luck has a larger role in the second case than in the first (effectively by a multiple of 4). In being treated by a non-specialist you have placed a far greater reliance on luck.

But consider further. You made your choice of a physician more or less at random within a medical establishment while nevertheless only 10% of the physicians are specialists of the relevant sort. But as it happened, you picked a specialist. How lucky was that?

Display 4.6

LUCK IN RECOVERING FROM A MALADY

Time Period (months)	% of Patients Recovered by this Time	Gain (months saved from possible hospitalization)	Expectation Rehospitalization	λ (amount)	λ^- (extent)
1	20%	4	2.5 months	+1.5	1
2	50%	3		+0.5	$\frac{3}{4}$
3	80%	2		−0.5	$\frac{2}{4}$
4	90%	1		−1.5	$\frac{1}{4}$
5	100%	0		−2.5	0

If your choice of physicians was indeed made at random, then there are two possibilities

Your survival with a specialist's treatment—probability: $.1 \times .95 = .095 = 9.5\%$
Your survival with a non-specialist's treatment—probability: $.9 \times .8 = .72 = 72\%$

With an overall survival probability of 81.5% under random choice while with an assured specialist it is 95% this means that the choice of an expert increases your chance by $(95 - 81.5) \div 81.5$ or 17%. And this mirrors the extent of your good luck in having chosen an expert.

A different medical example is encapsulated in the recovery-rate situation illustrated in Display 4.6, where luck is a matter of speedy recovery. Since some 75% of the patients fall into the λ-positive range, the situation is luck-favorable despite its negative mien. (Luck is not outcome-correlative but attentive-comparative!)

* * *

For yet another instructive case, consider the graduating high-school student who is pursuing college admission. She applies to institutions of different levels of accessibility, ranking them as per Display 4.7.

- a "reach" where acceptance and ultimate conclusion is very unlikely.
- a "realistic prospect" where admission is probable.
- a "fall back" where admission is virtually certain.

Display 4.7

A COLLEGE ADMISSION EXAMPLE

Outcome (O_i)	Admission Probability p_i	Admission Desirability $\lvert O_i \rvert$ [scale of 10]	Expectation E	λ	λ^-
O_1: "reach" admission	.1	10		3.9	1
O_2: "realistic" admission	.7	7	6.1	.9	$\frac{2}{3}$
O_3: "fall-back" admission	.2	1		−5.1	0

Our candidate here is in the relatively favorable position in that θ^+, the aggregate likelihood of a lucky result, is rather high at 80%. On the other hand the unluck of a "full-book" substantially exceeds the luck of a "reach."

* * *

Let it be that by 1980 statistics the life expectancy of Ruritanians born in 1880 was as per Display 4.8. And now let us ask: If Sigismund, a Ruritanian born in that year lived to 70 years of age, how lucky was he? The answer is provided in the tabulation. With E here at some 52 years the luck of such an individual is 70 – 52 = 18 years in amount. And some 30% of Ruritanians are fortunate in achieving as least this level.

Display 4.8

STATISTICS FOR RURITANIANS BORN IN 1870

Age-at-Death Cohort	Average Cohortal Lifespan	% of Cohortal Ruritanians	p	E	λ	λ^-
0-15	10	10	.10		−42	$\frac{0}{70} = 0$
15-30	20	5	.05		−32	$\frac{10}{70}$
30-40	35	15	.15	52	−12	$\frac{30}{70}$
40-60	55	25	.25		+3	$\frac{45}{70}$
60-75	65	35	.35		+13	$\frac{55}{70}$
75-	80	10	.10		+28	$\frac{70}{70} = 1$

NOTE: θ, the overall probability if achieving a lucky outcome is 70%.

* * *

Even a high-probability outcome can be very lucky when a low-probability alternative that it averts is catastrophic. And even an outcome that is very negative can be quite lucky when its alternatives are substantially worse. Someone who survives an earthquake's collapse of his apartment

Display 4.9

LUCK IN ONE-ROUND RUSSIAN ROULETTE WITH *n* BULLETS

Number (n) of bullets	Probability (p_n) of Survival	Survival Expectation E_n	Luck of Survival (1 – E_n)
1	$\frac{5}{6}$	$\frac{5}{6}V$	$\frac{1}{6}V$
2	$\frac{4}{6}$	$\frac{4}{6}V$	$\frac{1}{3}V$
3	$\frac{3}{6}$	$\frac{3}{6}V$	$\frac{1}{2}V$
4	$\frac{2}{6}$	$\frac{2}{6}V$	$\frac{2}{3}V$
5	$\frac{1}{6}$	$\frac{1}{6}V$	$\frac{5}{6}V$

Key:

n = number of bullets in a 6-chambered revolver

p_n = probability of survival (in the case of n bullets)

E_n = expectation of survival (in the case of n bullets)

λ_n = luck of survival (in the case of n bullets)

V = value of the agent's life

NOTE: With no bullets, survival is certain and so luck is absent. With 6 bullets the luck of survival is a non-issue—there is no survival. Either way, luck does not come into it.

building has every right to deem themselves very lucky to escape with a few scrapes and bruises.

* * *

Consider Russian Roulette, as per Display 4.9. Basic to the situation are these suppositions:

(1) *A* six shooter revolver with n bullets
(2) V = value of life
(3) p = probability of survival

So with n bullets we have the displayed situation.

Accordingly note that: When there are n bullets and the agent survives, the individual's need for good luck (λ) is exactly proportionate to the number of bullets (n). However, with 0 or with 6 bullets the result is certain: luck does not come into it.

Display 4.10

A COMPLEX SITUATION

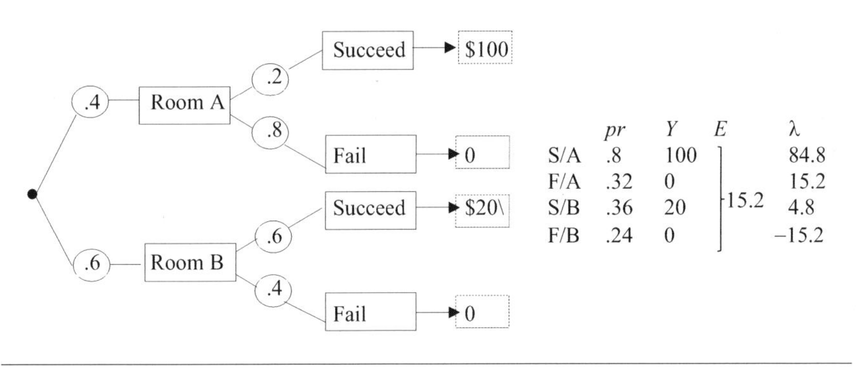

* * *

Let it be that—as per Display 4.10— a treasure-hunt participant has a 40% chance of being assigned to Room *A*, where he has a 20% chance of finding $100, in contrast with a 60% chance of being assigned to room *B* where he has a 60% chance of finding $20. How lucky is he if assigned to Room *A*?

Here we have:

Outcome	Luck (λ)
• Success in Room *A*	+$74.80
• Failure in Room *A*	−$15.20
• Success in Room *B*	+$4.80
• Failure in Room *B*	−$15.20

Any success is better than failure, but only in Room *A* is success particularly lucky.

Note, moreover, that if assigned to Room *A* our protagonist's expectation is $20, while if assigned to Room *B* it is $12. So the expectation of the initial assignment is .4(20) + .6(12) = 15.20. On this basis we have

$$\lambda(\text{Room } A \text{ assignment}) = 20 - 15.20 = 4.80$$

$$\lambda(\text{Room } B \text{ assignment}) = 12 - 15.20 = 3.20$$

And so the here is the answer to our initial question. The luck of a Room A assignment is \$4.80. And this is \$1.20 greater that the luck of a Room B assignment. The luck-favorite has always been: Room A. But we now have an answer to the question: By just how much?

* * *

Let it be that in the course of a trip you made a special detour to a restaurant noted for its excellent steaks, Arriving at 10 PM, you found that this was the only day of the week that the place does not close at 9 PM. You were certainly lucky. But how much so? For an answer we turn to the Basic Luck Equation $\lambda = Y - E$. The yield is one excellent steak dinner ($= S$). With a one out of seven chance of hitting the right day your success likelihood is one-seventh. The expectation in thus $\frac{1}{7} \times S + \frac{6}{7} \times 0 = \frac{1}{7}S$. On this basis we have $\lambda = S - \frac{1}{7} \times S = \frac{6}{7}S$. Your luck comes to *almost* one excellent steak dinner. By contrast had you arrived on another day your bad luck would have been $0 - \frac{1}{7} \times S = -\frac{6}{7}S$. Your bad luck here would *almost* have amounted to the loss of one excellent steak dinner.

Good luck is even possible in circumstances of economic loss and misfortune. If the market the widgets your company makes malfunctions and declines by twenty percent whence you had earlier managed to sell your thousand-dollar crop for nine-hundred, you sustained only a hundred dollar loss where one of two hundred was to be expected. The Basic Luck Equation $\lambda = Y - E$ now gives λ as $-100 - (-200) = \$100$, and thereby exactly captures the matter.

* * *

When thinking about luck one often has in mind something like a lottery when a substantial gain has unexpectedly been realized. But the other side of the matter—the chance avoidance of a great negativity also provides a vivid illustration. Luck dwells in the discrepancy between attainment and expectation.

Success against the odds is the quintessence of good luck. What makes a "lucky guess" lucky is exactly its unlikelihood. Thus consider: If by sheer happenstance someone picks the right answer out of 100 possibilities, then with a score of 1 for right and 0 for wrong, their expectation of

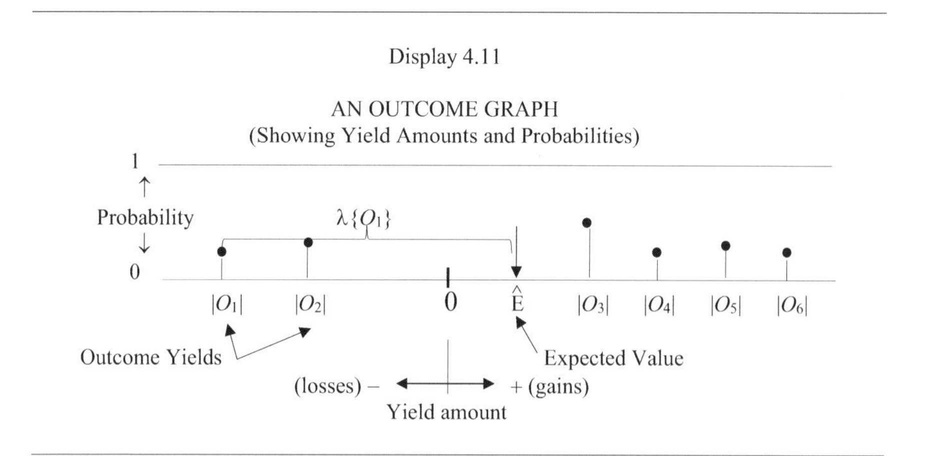

success stands as $E = \frac{1}{100} \times 1 + \frac{99}{100} \times 0 = .01$. So with luck subject to $\lambda\{O\} = Y - E$, we then have it that success luck stands at $1 - .01 = .99$. In the circumstance it cannot get much better than that. Analogously, the bridegroom is called "the lucky man" subject to the complimentary fiction that the bride is a pearl of rare price which he has carried off in the face of extensive competition.

Luck has a significant part to play even in those unfortunate chancy situations where no positive outcomes are on offer. For a negative outcome can prove to be very lucky in averting an alternative that is yet worse.

In luck deliberations the idea of a good outcome is equivocal. For it could mean either "an outcome of positive *yield* $|O| > 0$" or "an outcome of positive *luck* where $\lambda\{O\} > 0$." And these two are nowise equivalent, since a negative outcome can bring positive luck when all the alternatives are worse.

* * *

Display 4.11 depicts what might be called the *Outcome Graph* of a chancy situation, with its clear indication that luck reflects the distance between $|O|$ and E so that the worst outcome has negative luck and the best outcome positive, irrespective of how positive or negative those outcomes may be in themselves.

* * *

Consider a paradigmatic instance of a bad-luck situation, the case of inconvenient timing. Thus suppose that Smith occasionally misplaces his car keys, but in the present occasion is particularly unlucky as doing so when he needs to catch a flight to an important meeting. Here we have a particularly unlucky eventuation. How is this captured by the present approach?

The situation is exhibited in the following illustration:

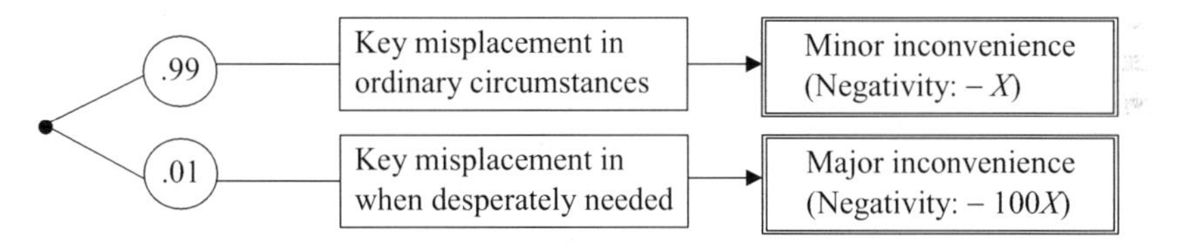

Here the expectation is $.99(-X) + .01(-100X) = -.99X - 1X \cong -2X$. And accordingly we have it via $\lambda = Y - E$ that:

$$\lambda(\text{ordinary circumstance}) = -X - (2X) = X$$

$$\lambda(\text{extraordinary circumstance}) = -100X - (-2X) = -98X$$

Clearly the bad luck of the second case is vastly greater since here the outcome is far worse than the normal expectation. And a comparable accounting could be given for situations answering to the familiar idea of "lucking out."

* * *

As the various examples considered here serve to illustrate, the assessment of luck in chancy situations can render instructive service in matters of rational deliberation and decision. In particular, whenever chancy situations are such that the realization of a positive outcome involves a good deal of luck, we have a red flag indicating the need for due care and caution.

Chapter 5
Luck in Success/Failure Situations

It is sensible to treat both win-or-lose competitions and find-or-miss searches conjointly because their conceptual structure is uniform from a theoretical point of view. The contrast between success and failure is critical on both sides alike.

* * *

To illustrate how the luck-measure $\lambda = Y - E$ works out in binary success/failure contexts consider win-luck situations of the following format:

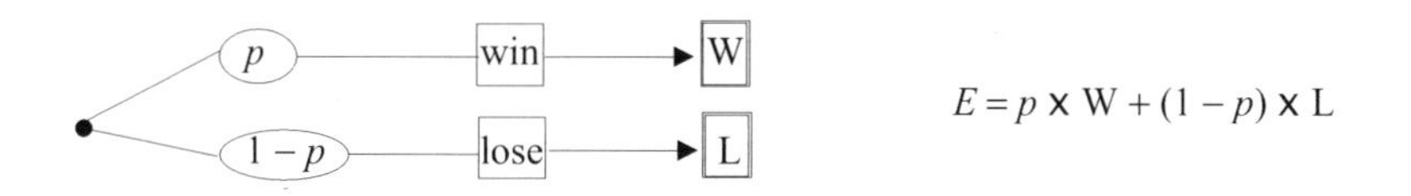

Here we have:

$$\lambda\{\text{win}\} = Y - E = W - p \times W - (1 - p) \times \text{L} = (1 - p)(W - L) = (1 - p) \times \Delta$$

with $\Delta = W - L$ being the stake at issue.

Success against the odds is optimally lucky. For in win-lose situations of this sort we have it that the luck of winning is basically a two-parameter issue:

$$\text{Win-luck} = (\text{win-improbability}) \times (\text{Stake})$$

Accordingly, the larger the stake and/or the smaller the probability of winning, the greater the luck of realizing success. It is always luckier than otherwise to win a "long shot," low-probability victory. And since the win-probability p can vary from 0 to 1, win-luck can vary from 0 to Δ, and lose-luck from $-\Delta$ to 0.)

Winning against the odds is the quintessence of luck. And with a fixed stake Δ luck comes to the probability at issue. Thus in sarcastically responding "Good luck"

N. Rescher, *Luck Theory*, Logic, Argumentation & Reasoning 20,
https://doi.org/10.1007/978-3-030-63780-4_5

to someone's adoption of problematic aims, we are sarcastically indicating that our appraisal of the likelihood of its realization is decidedly low.

* * *

Consider the situation of a 1/0 scored binary win/lose situation:

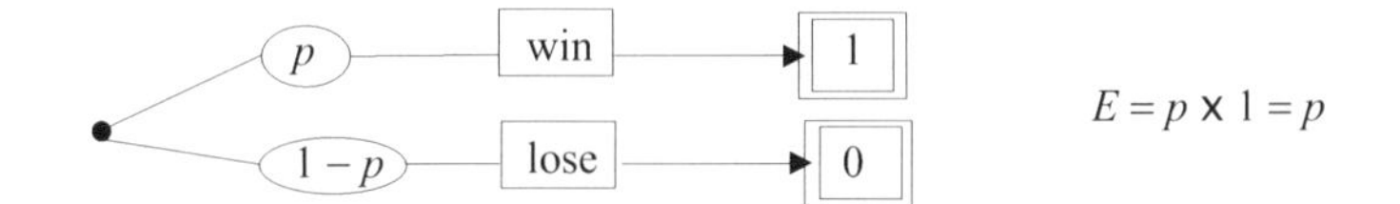

So here we have:

$$\lambda\{\text{win}\} = 1 - p$$

The bigger your win-probability the less your reliance on luck. And the relationship is linear. Increase p by $Z\%$ and your luck-reliance decreases by exactly that percentage. And of course it would make no sense to speak of "winning" or "losing" in situations where the outcome makes no difference.

* * *

In binary win/loss situations we will of course have it that

$$\lambda\{\text{win}\} = |\text{win}| - E = 1 - p$$
$$\lambda\{\text{lose}\} = |\text{lose}| - E = -p$$

Here the amount of win-luck is always the improbability, and of loss luck the negative of the probability.

* * *

In all binary win/lose situates the various luck variants of win-luck have uniform value:

$$\lambda^{+}\{\text{win}\} = 1$$
$$\lambda^{-}\{\text{win}\} = 0$$
$$\lambda^{\#}\{\text{win}\} = 0$$

This means that it could only ever happen that $\lambda\{\text{win}\} = \lambda\{\text{lose}\}$ when $|\text{win}| = |\text{lose}|$. Wins and losses are indifferent only when nothing is at stake.

In fair binary contests where $E = 0$ the good luck of one party is always balanced off by the bad luck of another. For where X's win always correlate with Y's loss we have it that for any outcome O: $|O|_x = -|O|_y$. And with $E_x = E_y = 0$ we have it that $\lambda_x\{O\} = |O|_x - 0 = -|O|_y - 0 = -\lambda_y\{O\}$.

One is lucky in winning. But how much luckier is it to win twice when the chancy situation repeats? Intuition says: "Twice as lucky." What does mathematics say?

Consider a binary win/lose gamble with a win-probability p. The expectation for a single winning round of this gamble is $\lambda\{W\} = |W| - E$. But with a double win the yield and the expectations both double. And the luck of a double win is $2 \times |W| - 2 \times E$ or $2 \times \lambda\{\mathrm{W}\}$, so that doubling the win doubles the luck. Happily intuition and mathematics agree.

* * *

On the basis of their performance record, A and B are evenly matched at tennis, each with a 50% chance of winning a set against the other. This time, however, A bests B in two straight sets. How lucky was this for A?

Consider A's situation:

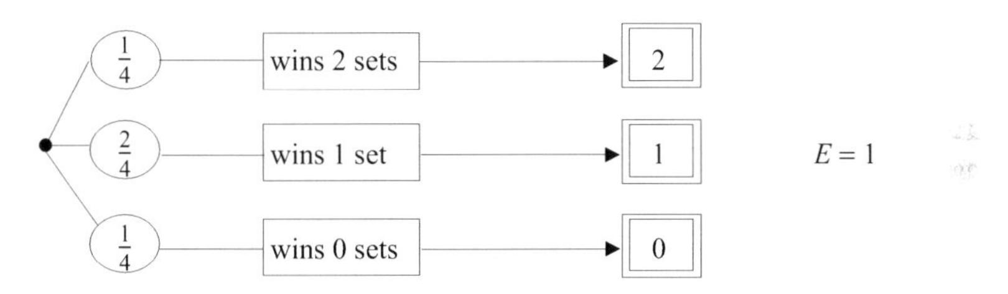

Via $\lambda = |O| - E$ we have it that:

$$\lambda\{\text{wins 2 sets}\} = 2 - 1 = 1$$

$$\lambda\{\text{wins 1 sets}\} = 1 - 1 = 0$$

$$\lambda\{\text{wins 0 sets}\} = 0 - 1 = -1$$

Thanks to their even matching a one set result will have the luckless upshot of "being only what is to be expected." And the luck of A's two-set win is exactly equal and opposite that of a two-set loss.

* * *

Just when would the luck in a chancy situation be equal and opposite? Consider:

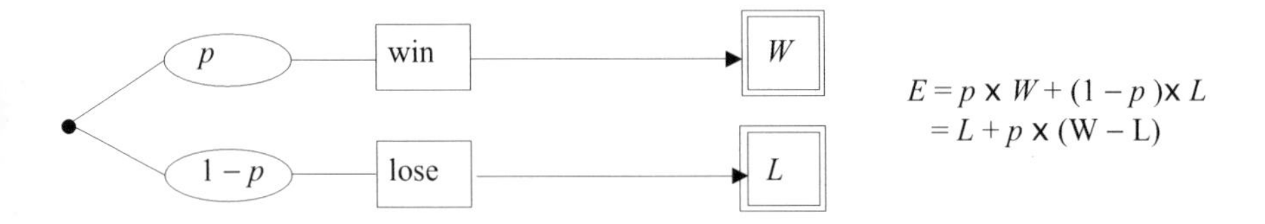

Here we have

$$\lambda\{\text{win}\} = W - E = W - [L + p(W - L)] = (1 - p) \times (W - L)$$
$$\lambda\{\text{lose}\} = L - E = L - [L + p(W - L)] = p \times (L - W)$$

So $\lambda\{\text{win}\} = -\lambda\{\text{lose}\}$ as impossible unless $W = L$. And $\lambda\{\text{win}\} = -\lambda\{\text{lose}\}$ iff $p - \frac{1}{2}$, regardless of the size of W and L.

* * *

It is instructive to consider how λ-luck changes in binary choice situations over the range of variation of E and p. We thus begin with

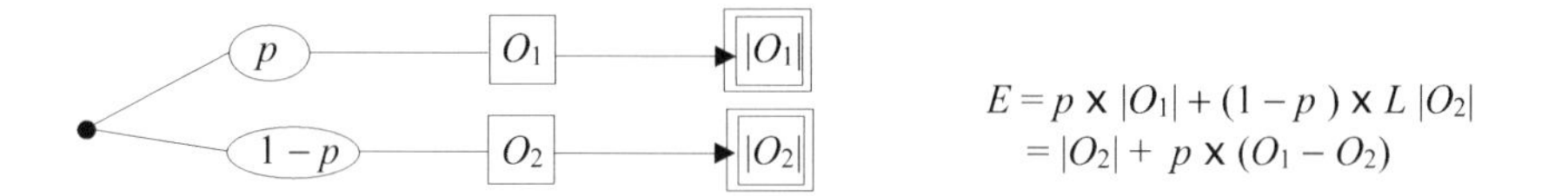

We now have:

$$\lambda\{O_1\} = |O_1| - E = |O_1| - |O_2| - p \times (O_1 - O_2) = (1 - p) \times (O_1 - O_2)$$
$$\lambda\{O_2\} = |O_2| - E = |O_2| - |O_2| - p \times \Delta = -p \times (O_1 - O_2)$$

The behavior of win-luck in point of outcome variation is illustrated in Display 5.1. Significantly, the 100-fold increase of Z from 10 to 1000 changes the polarity of $\lambda\{O\}$ from positive to negative.

* * *

Let someone be in the coin-toss where if the coin comes up Heads then they will win \$1,000, and if it comes up Tails they are to get the same

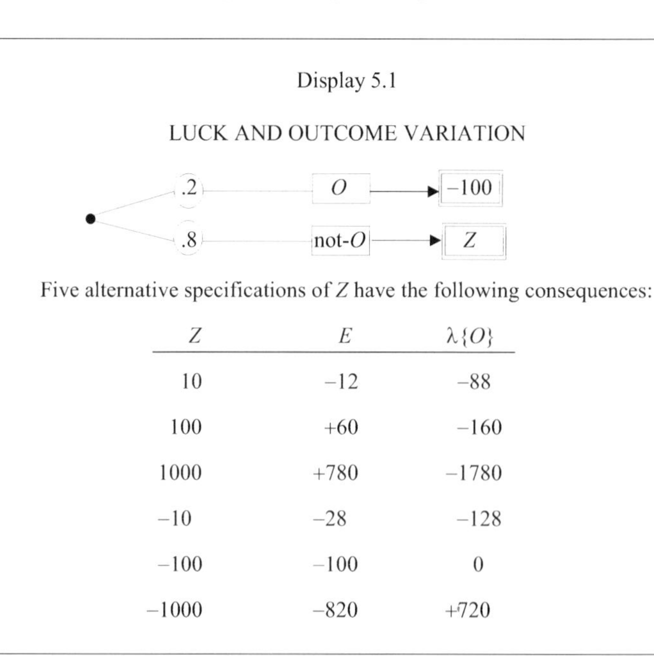

Display 5.1

LUCK AND OUTCOME VARIATION

Five alternative specifications of Z have the following consequences:

Z	E	$\lambda\{O\}$
10	−12	−88
100	+60	−160
1000	+780	−1780
−10	−28	−128
−100	−100	0
−1000	−820	+720

opportunity again—and thus have the same expectation. Then their initial expectation is such that we have

$$E = \frac{1}{2} \times 1,000 + \frac{1}{2} \times E \text{ or equivalently } E = 100$$

So if Heads comes up (whenever that may eventually be) then on the basis of $\lambda = Y - E$ the win-luck will be:

$$\lambda\{\text{win}\} = 1,000 - 1,000 = 0$$

With ultimate success effectively certain, our protagonist will be no luckier a win in the first round than in the second—or the 100th. The point of course is that ultimately winning 1,000 is effectively inevitable and this certainty precludes luck. (Unless, of course, waiting time is figured negatively into the yield-value.)

The influence of probabilistic variation upon luck is illustrated in Display 5.2. In the postulated circumstances, irrespective of probability p gaining that \$100 is never unlucky and losing \$500 is never lucky. But the *amount* of luck is much dependent on the value of that probability, diminishing as that probability increases and ultimately vanishing when that result is a *fait accompli.*

Display 5.2

LUCK AND PROBABILITY VARIATION

p	O	→	+100
$1-p$	not-O	→	−500

$E = 100p + (1 - p) \times -500$
$= -500 + 600p$

Five alternatives for p:

p	E	$\lambda\{O\}$	$\lambda\{\text{not-}O\}$
.1	−440	+540	−60
.3	−320	+420	−180
.5	−200	+300	−300
.8	−20	+120	−480
1.0	+100	0	−600

---

* * *

This failure to find luck in an assured gain may seem counterintuitive. To come to terms this reaction one must once move to ponder the distinction between good fortune and luck.

You make a gamble and succeed. You make it once again, and again you succeed. How much luckier are you with that double success that you were with the initial single? Intuition says twice. Does calculation agree? Let's see.

Consider a simple win-lose gamble as per:

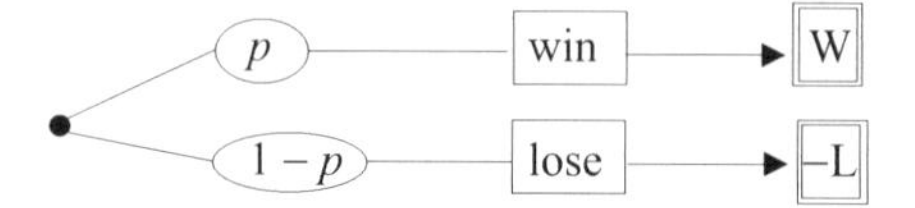

Here $E_1 = p \times W + (1 - p) \times L$. And accordingly $\lambda\{\text{win}\} = W - E_1 = (1 - p) \times (W - L)$.

5.1 Searching

Searches are a fertile domain for deliberation about luck. If you do not misplace the TV remote, you need not hunt for it, and so are not at luck's mercy for finding. Searching as such is not always a problematic endeavor. By and large one is fortunate to avoid it when possible.

There is, of course, such a thing as *serendipity*, the phenomenon of finding without searching—of making findings or discoveries in unplanned, accidental encounters rather than by actual effort. When such a finding is of significant interest or value, then the luck involved is substantial, with high yield and low expectation, so that $\lambda = |O| - E$ is large.

* * *

That is a random canvass of equiprobable compartments a straightforward example of a search proceeding and illustrate various key principles of luck in searching. Thus with a random search for a single object in a space of N compartments the probability of immediate success is $\frac{1}{N}$. And in the event of failure here, the probability of success in the next step rises to $\frac{1}{N-1}$. And so on, until after $N - 1$ steps we arrive at 1. And of course as probable success grows, luck declines as less and less is required. (The general situation is depicted in Display 5.3)

How much luck is involved in achieving success when N or $100N$ alternatives must be explored? When a multilateral investigation splits the terrain into N compartments each assigned to a particular investigator, and it happens that the sought-for resolution falls into X's compartment, does the credit belong to X (who gets this compartment by luck) or to the team as a whole? It credit greater for a discovery made by method rather than sheer luck? The whole domain of search and research is pervaded by instructive luck issues.

Display 5.3

LUCK IN SEARCHING
(Success in a Random Search of N Components)

N	$p = \frac{1}{N}$	E	$\lambda\{\text{success}\} = 1 - E$ in %	λ^+
2	$\frac{1}{2}$	$\frac{1}{2}$	50%	.5
10	.1	.1	90%	.9
100	.01	.01	99%	.99
1000	.001	.001	~100%	~1
10,000	.0001	.0001	~100%	~1

* * *

In searching, added information can readily diminish the scope for luck. Suppose one is trying to locate something within a tic-tac-toe grid, where it is known from the outset that it is not in the center position. Beyond this, it's all guesswork, all chance. In striving to guess aright, one will clearly place a great deal of reliance on luck. (To be precise $\lambda\{\text{success}\} = 1 - \frac{1}{8} = \frac{7}{8}$.) But now suppose being further informed that the objective is not in the first row. At once the probability of success p increases by 7.5% from $\frac{1}{8}$ to $\frac{1}{5}$, and so E changes to $\frac{1}{5}$ as well. In consequence, the λ luck of a correct guess, here $1 - p$, reduces to $\frac{4}{5}$, so that your dependence on luck decreases from $\frac{7}{8} = \frac{35}{40}$ to $\frac{4}{5} = \frac{32}{40}$, also declining by $\frac{3}{40}$ or 7.5%.

To be sure, it may seem anomalous that while the range of uncertainty has decreased from 8 to 5—i. e. by some 37%—the correlative luck-of-finding has decreased only by 7.5%. But this perspective is misleading.

* * *

Searches vividly exhibit the perspectival duality of luck. If you have managed to find the needle in the haystack you are very lucky. But if you are so positioned that you *need* to have it, and actually require all that luck, then you are decidedly unfortunate. The irony of luck lies in the fact that it is fortunate not to need luck. For if reaching goal G requires the luck at issue, $|G| \times (1 - p)$, to be substantial, and thus the probability of success to be well below 1, then you are not in a particularly fortunate situation.

Chapter 6
Basic Luck Theorems

A great many characteristic features of luck are readily established as demonstrable theorems subject to the understanding that the luck of chancy situations can be assessed quantitatively. This can be done on basis of the Basic Luck Equation to the effect that the quantity of an outcome's, luck is the difference between its yield and its situation's expectation:

$\lambda = Y - E$ or more fully $\lambda\{O\} = |O| - E$.

The following theorems implement this idea.

Theorem I *Be it positive or negative, any outcome in a chancy situation whose yield exceeds the expectation is bound to qualify as lucky.*

Proof An outcome O is lucky when $\lambda\{O\} > 0$. But since $\lambda\{O\} = |O| - E$, the theorem immediately follows when $|O| > E$.

Theorem II *With negative expectations the level of an outcome can exceed its yield.*

Proof This is an instantive consequences of λ's depictions.

Theorem III *Maximal luck always attends the optimal outcome.* (Of course it need not be positive.)

Proof Since $\lambda\{O\} = |O| - E$, with E a constant for any chancy situation at issue, it is clear that when $|O|$ is maximal, $\lambda\{O\}$ must be so as well.

Theorem IV *Luck is contrastive. It can never happen in chancy situations that* every *possible outcome is luck or unlucky.*

Proof Given that $\lambda = Y - E$, it is not possible for every outcome to be super or again sub-expectational. For by the definition of E this cannot happen.

Theorem V *With inevitability there is no luck. When an outcome is effectively inevitable (i.e., when $pr\{O_i\} \approx 1$, so that $pr\{O_j\} \approx 0$ for all $j \neq i$), then $\lambda\{O_i\} = 0$.*

N. Rescher, *Luck Theory*, Logic, Argumentation & Reasoning 20,
https://doi.org/10.1007/978-3-030-63780-4_6

Inescapable outcomes, no matter how unfortunate, involves zero luck, seeing that the essential factor of chanciness being absent.

Proof This follows straightway from $\lambda\{O_i\} = |O_i| - E$, which will be 0 since now $|O_i| = E$.

Corollary When all outcome yields are equal, the totality of luck is Zero.

Proof $E = \sum [(|O_i| \times p_i = \sum (|O| \times p_i) = |O| \times \sum p_i = |O|$. Therefore here $\lambda\{O_i\} = |O| - \sum = |O| - |O| = O$.

Theorem VI *Disappointed expectations typify bad luck.*

Proof With luck assessed as per $\lambda = |O| - E$, luck will clearly decrease with an increase in E. Missing out as a benefit when the probabilities warrant greater expectations is the quintessence of bad luck.

Theorem VII *Luck is contextual: The luck of an outcome depends on what its situational alternatives are.*

Proof Luck as per $\lambda\{O\} = |O| - \mathrm{E}$ obviously does not depend on $|O|$ alone (and not even with its probability considered). It varies with E and then depends on the alternatives and their likelihoods.

Corollary The luck afforded by a particular outcome depends not just on its probability but upon the structure of its expectation context.

Proof Consider

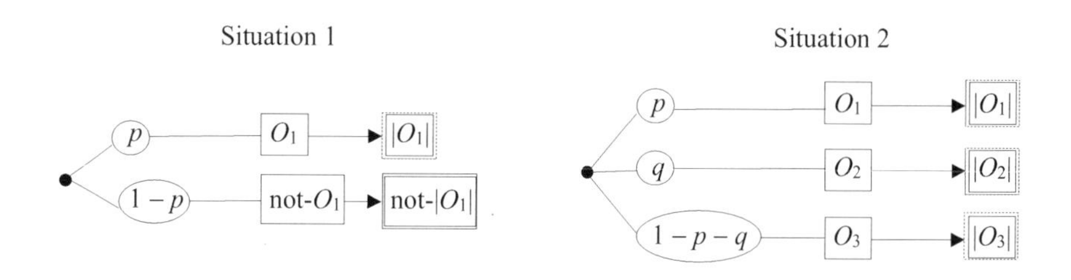

Here:

$$E_1 = p \times |O_1| + 1 - p \times |\text{not-}O_1| E_2 = p \times |O_1| + q \times |O_1| + (1 - p - q) \times |O_3|$$

$$\lambda_1\{O_1\} = |O_1| - E_1 \lambda_2\{O_1\} = |O_1| - \lambda_1\{O_1\} = |O_1| - E_1$$

For these to be the same we would need to have $E_1 = E_2$ or equivalently

$$(1 - p) - |\text{not-}O_1| = q \times |O_2| + (1 - p - q) \times |O_3|$$

But there are no considerations of general principle to ensure that this relationship will obtain.

Corollary *Luck is counterfactual.* The luck of an outcome would be different if the available alternatives were different, and thereby depends on what might or would have happened if the contextual situation had not been as is.

Theorem VIII *Impredictability is a necessary but not sufficient condition for an outcome's being lucky.*

Proof of Necessity If the outcome is securely predictable—i.e., has probability 1—then its expectation equals its yield. Hence $\lambda = Y - E = 0$.

Proof of Insufficiency Consider

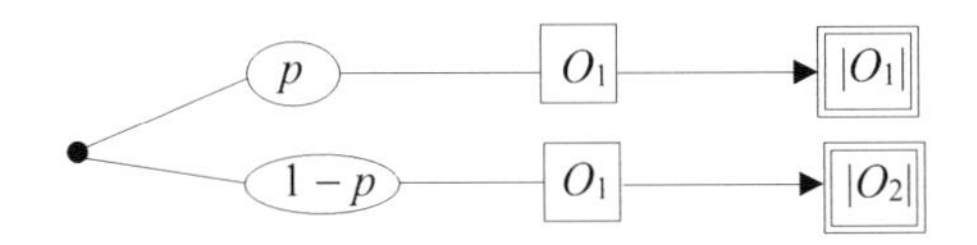

New $p \neq 0$ does not ensure that $\lambda\{O_i\} \neq 0$. For this last comes to $E \neq Y$, and $p \neq 0$ does not ensure this.

Theorem IX *In chancy situations any outcome that averts a catastrophe invariably qualifies as very lucky.*

Proof Consider a situation of the relevant format:

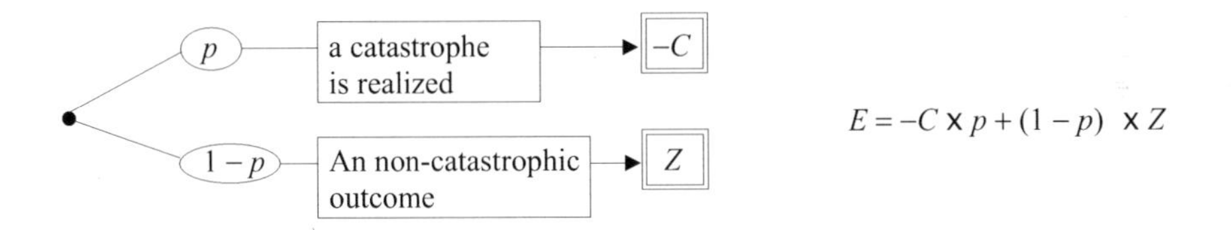

Here $\lambda\{\text{non} = \text{catastrophe}\} = Z - E = Z + C \times p - (1 - p) \times Z = C \times p - p \times Z = p(C + Z)$. Since C is very big in relation to Z, this is bound to be large.

Theorem X *There is a disconnect between good luck and benefit (and between bad luck and loss). Lucky outcomes need not be positively beneficial, nor unlucky ones disadvantageous.*

Proof Clearly, even positive outcomes can fall far short of available optimality, and negative ones can avert yet greater negativities. (A bad outcome is still lucky when all the alternatives are yet worse.)

Theorem XI *In any binary win-or-lose situation the difference between win-luck and lose-luck always exactly equals the stake.*

Proof In a binary gamble the size of the stake (Δ) is measures by

$$\Delta = |\text{win}| - |\text{lose}|.$$

But since $\lambda = \text{Y} - E$ one has it that: $\lambda\{\text{win}\} = |\text{win}| - E$, and moreover $\lambda\{\text{lose}\} = |\text{lose}| - E$. And so the desired conclusion

$$\Delta = \lambda\{\text{win}\} - \lambda\{\text{lose}\}$$

at once follows.

Theorem XII *In a 0/1 scored binary win-or-lose situation the win-luck is the product of the probability of losing and the loss-luck the negative of the win probability.*

Proof This follows directly from the definition at issue.

Theorem XIII *In a +1/0/−1 scored win-lose-draw situations the expectation's negative equals the luck of a draw.*

Proof With $\lambda\{\text{draw}\} = |\text{draw}| - \text{E}$ and $|\text{draw}| = 0$ the theorist follows.

Theorem XIV *Quantitative λ-luck as measured by* $\lambda\{O\} = |O| - E$ *can range from* $-\infty$ *to* $+\infty$. *But comparative λ*-luck as assessed by* $\lambda^*\{O\} = \frac{|O|-|O^-|}{|O^+|-|O^-|}$ *ranges from 0 to 1.*

Proof The first assertion follows from the indeterminate range of $|O|$. The second assertion follows from the fact that $|O|$ lies between $|O^-|$ and $|O^+|$.

Theorem XV *In a +1/0/−1 scored win-draw-lose situation, the λ*-luck of winning is +1 and of losing—0; but that of a draw is* $\frac{1}{2}$.

Proof When $|\text{win}| = +1$, $|\text{draw}| = 0$, and $|\text{lose}| = -1$, then all this follows at once from $\lambda^*(O) = \frac{|O|-|O^-|}{|O^+|-|O^-|}$

Theorem XVI *All narrow escapes are very lucky. They will have a large positive λ-value and a λ*-value close to 1.*

Proof For there to be a narrow escape, not only must some possible outcome be highly negative, but what actually happens must be substantially better what one would otherwise expect. Accordingly $|O|$ must be substantially bigger than E. But since $\lambda\{O\} = |O| - E$ this quality must be large. Moreover, $|O|$ must now be very much larger than $|O^-|$, so that $|O| - |O^-| = (\text{big})$. This puts $|O| - |O^-|$ close to $|O^+| - |O^-|$ so that presumably the ratio $\lambda^*\{O\} = \frac{|O|-|O^-|}{|O^+|-|O^-|}$ is close to 1.

Theorem XVII [The Fundamental Theorem of Luck.] *In any chancy-outcome situation the potential for good and bad luck balances out in the sense that the overall expectation of luck is always zero.*

Proof Since the luck associated specifically with outcome O_i is $\lambda\{O_i\} = |O_i| - E$, the overall expectation of luck is thus $\sum_{i}^{n} [(|O_i| - E) \times p_i]$. But we have:

$$\sum_{i}[(|O_i| - E) \times p_i] = \sum_{i}(|O_i| \times p_i) - \sum_{i}(E \times p_i)$$

$$= \sum_{i}(|O_i| \times p_i) - E = E - E = 0$$

The totality of the luck expectable over the manifold of possible outcomes is always process. The distribution of luck over outcomes is, in the end, a zero-sum game: greater luck for some outcomes must always balance off against lesser luck for others.

Theorem XVIII *Luck is reflexive. In chancy situations the luck inherent in having a certain amount of luck—the luck of being lucky to a certain extent—is just exactly that amount of luck itself. In effect, being lucky is itself lucky.*

Proof From $\lambda = Y - E$ it follows that the luck of outcome O comes to $\lambda\{\lambda\{O\}\} = |\lambda\{O\}| - E_\lambda$. But since $E_\lambda = 0$ by Theorem XVII, this is simply $\lambda\{O\}$ itself.

Theorem XIX *Limits to a particular outcome's luck in any chancy-outcome situation are set by its yield in relation to the alternatives, as per:*

$$|O_i| - |O_i^+| \leq \lambda\{O_i\} \leq |O_i| - |O_i^-|$$

Proof From $\lambda = Y - E$ it follows that since E is confined to the range between $|O_i^-|$ and $|O_i^+|$, so that $|O_i^-| \leq E \leq |O_i^+|$, $\lambda\{O_i\}$ will be confined to the range between $|O_i| - |O_i^+|$ and $|O_i| - |O_i^-|$. QED.

Theorem XX *In any win/tie/lose outcome situation with these outcomes scored as per +1|0|−1 we have:*

$$\lambda\{\text{win}\} = \lambda\{\text{tie}\} + 1 = \lambda\{\text{lose}\} + 2$$

Proof Seeing that

$$|\text{win}| = +1, |\text{tie}| = 0, |\text{lose}| = |-1|, \text{and } p\{\text{win}\} - p, p\{\text{tie}\} = 1 - p - q, \text{and } p\{\text{lose}\} = q.$$

we have:

$$E = p + 0 \times (1 - p - q) - q = p - q$$

It then follows from $\lambda\{O\} = |O| - E$ that:

$$\lambda\{\text{win}\} = +1 - (p - q)$$

$$\lambda\{\text{tie}\} = 0 - (p - q)$$

$$\lambda\{\text{lose}\} = -1 - (p - q)$$

So $\lambda\{\text{win}\} = \lambda\{\text{tie}\} + 1 = \lambda\{\text{lose}\} + 2$. QED.

Theorem XXI *Whenever the risk of loss is great in that there is a certain outcome O whose yield $|O|$ is massively negative, its realization is bound to be extremely unlucky unless the alternatives are yet worse.*

Proof The fact that $\lambda\{O\} = |O| - E$ renders luck quasi-proportional to yield within that chancy situation at issue.[1]

Theorem XXII *A course of action in pursuit of a goal may well put too much reliance upon luck.*

Proof Consider the situation where a highly valued good has small chance of realization, so that we have:

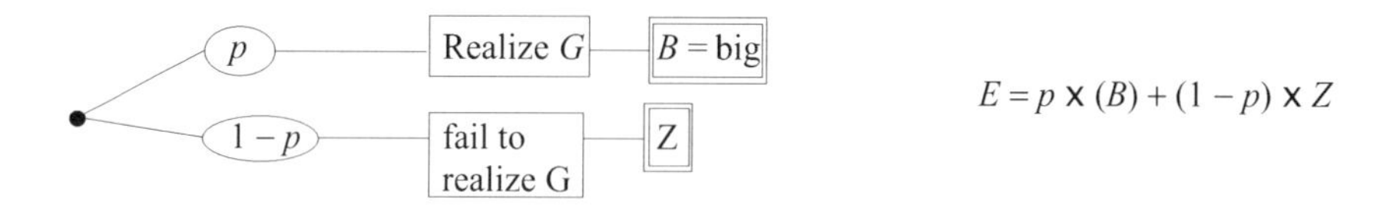

The luck at issue with general attainment will now be

$$\begin{aligned} \lambda = B - E = B - p \times B + (1 - p) \times Z \\ = B(1 - p) + (1 - p) \times Z \\ = (1 - p) \times (B + Z) \end{aligned}$$

In the circumstances $1 - p$ is bound to be close to 1 and $B + Z$, larger. So a great deal of luck is at issue with goal realization.

Theorem XXIII *The luck of a compound outcome is the probabilistically weighted average of the luck of its components.*

Proof To see how this must be so, compare:

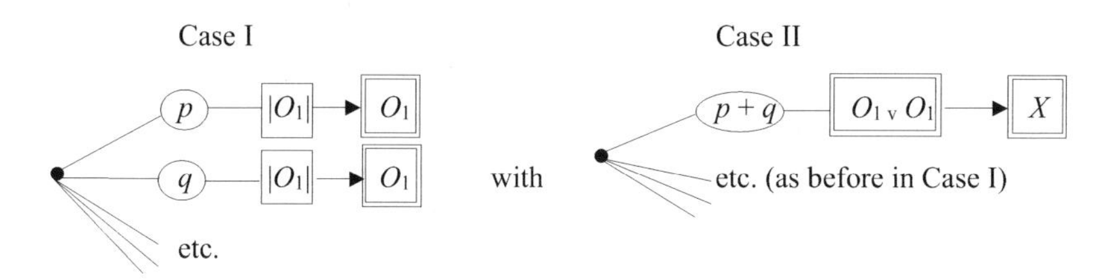

For the expectation to remain the same when Case I is amalgamated into Case II we must have it that

$$p \times |O_1| + q \times |O_2| = (p + q) \times X$$

where

$$X = \frac{p \times |O_1| + q \times |O_2|}{p + q}$$

Now since $\lambda = |O| - E$ we have

$$\lambda\{O_1 \vee O_2\} = X - E$$
$$= \frac{p \times |O_1| + q \times |O_2|}{p + q} - E$$

Accordingly:

$$\lambda\{O_1\} = |O_1| - E \text{ so that : } \quad |O_1| = \lambda\{O_1\} + E$$
$$\lambda\{O_2\} = |O_2| - E \text{ so that : } \quad |O_2| = \lambda\{O_2\} + E$$

With these substitutions in the just-given equation we obtain

$$\lambda\{O_1 \vee O_2\} = \frac{p \times (\lambda\{O_1\} + E) + q \times (\lambda\{O_2\} + E)}{p+q} - E$$
$$= \frac{p \times \lambda\{O_1\} + p \times E + q \times \lambda\{O_2\} + q \times E}{p+q} - E$$
$$= \frac{p \times \lambda\{O_1\} + q \times \lambda\{O_2\}}{p+q} \text{Q.E.D.}$$

Corollary *With equiprobable outcomes, the luck of a compound outcome is simply the average of that of its components.*

Theorem XXIV *The luck-profit of an outcome, O, namely is* $\mathcal{L}\{O\} = \lambda\{O\} - \lambda\{not - O\}$, *always equals* $|O| - |not$ - $O|$, *the yield differentiates between* |O| *and* $|not = O|$.

Proof Given that, $\mathcal{L}\{O\} = (|O| - E) - (|\text{not - } O| - E)$ so the desired conclusion at once follows. (So luck-profit is simply profit as such.)

6.1 Commentary

Central to these deliberations is Theorem XVII which deserves to be called the Fundamental Theorem of Luck. It maintains that in any chancy situation the (mathematical) *expectation* of luck—the sum total of the likelihood adjusted luck afforded by the various possible outcomes—is exactly zero. The good luck inherent in the prospect of positive outcomes is only possible at the price of the bad luck of others. The total outcome of yield may be immense and every available outcome may afford a positive benefit; but the situation as regards the prospect of luck is something very different: the overall *expectation* of luck in any chancy situation is nil.

* * *

The preceding deliberations serve to indicate that the quantitative *modus operandi* of luck is subject to exact reasoning of mathematical rigor. For the machinations of luck in chancy-outcome situations are clearly manifested in the interaction of the probabilities and yield-value of the possible outcomes. In view of this critical circumstance, the articulation of a mathematical theory of luck becomes a comparatively straightforward matter.

Note

1. $F(x)$ is quasi-proportional to $G(x)$ iff $F(x) = kG(x) + c$. Quasi-proportional parameters measure essentially the same factors.

Chapter 7
Luck and Risk

We speak of risk of capture in warfare or crime, of losing in games or fighting, of accident in transport, and so on. Risk is always a matter of potential loss or negativity. To be sure there is also the risk of "losing out"—in not realizing a positivity of some sort—but this too is in the end a sort of negativity.

One is at risk in any situation that has the possibility of issuing in a somehow negative result. (Only negative outcomes are at issue with risk: there is no such thing as a positive risk.) Given how easily matters can generally go awry most chancy situations are thus to some extent risky.

There are both risks we that knowingly undertake and risks that we are totally unaware of, risks we accept and risks we unwittingly run. In driving a car across town we run and accept the risk of accident. When the burglar who is recently released from prison prowls our neighborhood, we are in fact at risk of a visit from him, but blissfully ignorant of it. In both circumstances alike, if the risk passes us by, unharmed, then—know it or not—we are lucky, since either way the status quo is maintained in the face of a possible loss. To be sure, there may well be a chance that even situations of no possible loss one may fail to realize some unusually large benefit. But what this portends is not a risk but a lost opportunity.

Risks have two principal characteristic features: *magnitude* and *threat*. The *magnitude* of the risk depends on the severity or size of the negativity at issue; the *threat* of the risk depends on the likelihood of its realization. In the assessment of risk both factors are crucial.

* * *

Risk—the potential for loss—will be absent in situations where every outcome has a positive yield. However, situations where all outcomes are negatively yield-identical will be risky, despite necessity being entirely luckless. Here risk will not be absent—but is actually unavoidable.

N. Rescher, *Luck Theory*, Logic, Argumentation & Reasoning 20,
https://doi.org/10.1007/978-3-030-63780-4_7

Display 7.1

RISK ASSESSMENT FACTORS

$\lvert O^{-}\rvert$	assesses *gross risk* and (when negative) answers the question: What is the *amount* of the worst-case loss?
$\lvert O^{-}\rvert \times pr\{O^{-}\}$	assesses *net risk* and (when negative) and answers the question: What is the *prospect* of the worst-case loss?
E	assesses the *expectation* and (when negative) answers the question: To what extent is the expectation unfavorable? Note: $E = \sum_{i} (\lvert O_i\rvert \times pr\{O_i\})$
E^{-}	assesses the *negative expectation* and answers the question: What is the sum total of the negative yield-moments. Note: $E^{-} = \sum_{i} (\lvert O_i\rvert \times pr\{O_i\})$for all and only negative $\lvert O_i\rvert$ (Analogously with E^{+} for the *positive expectation.*)
θ^{-}	assesses the loss probability and answers the question: What is the total probability of having a negative yield-outcome. Note: $\theta^{-} = \sum_{i} \lvert O_i\rvert$ for all and only negative $\lvert O_i\rvert$

* * *

In caring about risk we need to focus on the downside of chancy situations. Our concern should be with questions on the order of those canvassed in Display 7.1. These are the crucial *parameters of risk* that should condition or view of the acceptability of chancy situations. Taken in combination these give a rounded picture of the overall loss-potential in chancy situations.

The worst-case outcome $\lvert O^{-}\rvert$ provides—when negative—one of the prime indices of risk. With the Luck Equation having it that $\lambda\{O^{-}\} = [O^{-}\rvert - E$, this gives luck a leading role in risk assessment. For $\lambda\{O^{-}\} + E$ is a significant risk indication as well.

* * *

Regrettably, however, there is no one single decisive measure of unacceptable risk. Risk is a sufficiently complex and multi-faceted phenomenon for which no single measure is able to supply one-size-fits-all standard. A reasonable assessment of the extent of risk requires taking all of its various facets conjointly into appropriate account.

* * *

Consider the *outcome profile* for the following risky situation with six outcome possibilities indicated as to yields and likelihoods by dots the six dot possibilities indicated with regard to yield and probability:

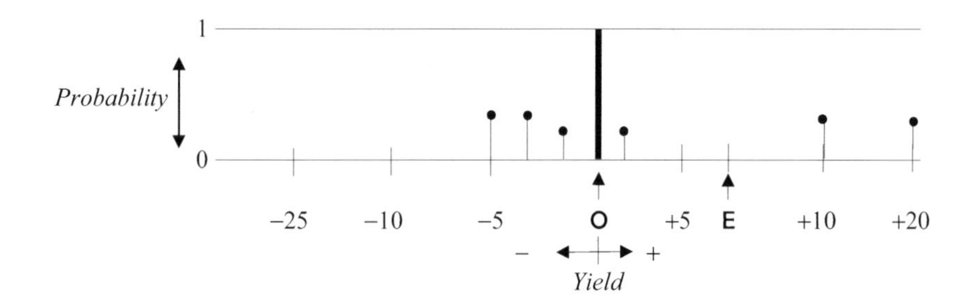

The outcomes to the right of the origin O are positive (gains) those to the left are negative (losses).

All of the risk parameters are derivable from the data of such a profile. The outcomes to the right of E are lucky in having their yield exceed the expectation; those to the left of E are unlucky. To be unlucky an outcome need not be a loss, nor need it be a gain to be lucky. The best of a bad lot can yet be lucky. Doing well is one thing, doing better than expected another. And so luck is not an inevitable comparison of risk. Neither need a risky (i.e. negative) outcome be unlucky (*E*-underachieving), nor must a lucky (E-overachieving) outcome be risky (i.e., negative).

* * *

One significant consideration regarding the acceptability of a chancy situation roots in the question: "Just how lucky is it to avert the worst-possible outcome?" The answer here lies in the value of $\lambda\{O^{-}\} = |O^{-}| - E$.

Consider here the following alternative yield distributions for four equiprobable outcomes.

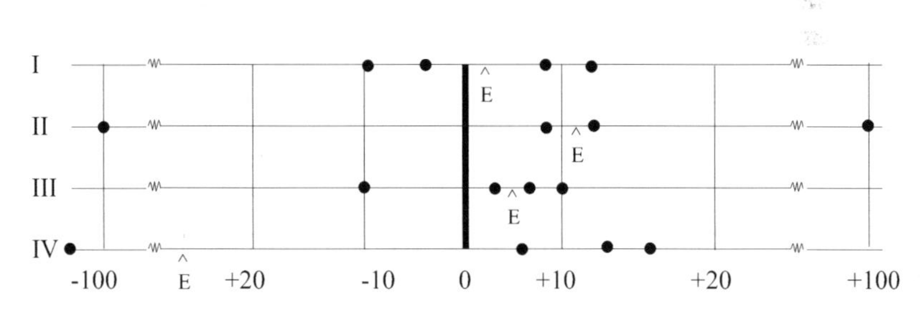

For these four situations we have $\lambda\{O^{-}\}$ as -12, -88, -5, and -100, respectively, with corresponding expectations of +3, +11, +5 and -35.

Accordingly, the risk-averse had best stick to alternatives I or III, notwithstanding their small expectation, while the risk accepting could well prefer II.

* * *

Accordingly luck and risk are substantially independent factors. Luck is a matter of how outcome relate to expectations (either way, positive or negative); risk a matter of how they relate to the origin (on its negative side). Thus even a no-risk situation can have lucky outcomes, and even a lucky outcome can obtain in a no-risk situation. A very unlucky outcome can occur in a no-risk situation (when that outcome, though positive, occurs in a situation where all others are far better.)

One significant connection between risk and luck is implicit in the fact that $|O^-|$ always the unluckiest of possible outcomes and (when negative) a prime indicator of risk. In general risk is entangled with luck, and usually the riskier an enterprise is, the more luck is required to bring it off. For since $\lambda\{O^-\} = |O^-| - E$ the gross risk $|O^-|$ will be $\lambda\{O^-\} + E$. This establishes a lock-step coordination between gross risk and the luck inherent in worst-case outcomes.

However, in principle risk and luck are detached issues. If all the possible outcomes are bunched together on the negative side of the value axis, there is no escape from risk because misfortune is inevitable. However if you realize the one and only loss that is far less than the rest, you are lucky indeed. all the outcomes are on the positive side there is yet room for luck (when some are far better that others) but no place for risk.

Success in a risky enterprise is the very paradigm of a good luck. For when an optimal outcome $O+$ is realized in conditions when the expectation E is minimal, the resultant luck $|O^+| - E$ will be as good as it gets.

* * *

Interesting questions arise regarding the relationship between luck and risk:

- *Question 1*: Can one be lucky in situations where nothing is at risk?

  Answer: Certainly. For when risk is absent because all possible outcomes have positive yields, still, if one of them is vastly greater than the rest, it will exceed the expectation so that its realization is indeed lucky.
- *Question 2*: Is it always lucky to avoid a risky (negative-yield) outcomes?

  Answer: Not necessarily. For suppose a chancy situation with 10 equiprobable outcomes, 5 with yields in the -10 to -12 range, one with $+10$, and 4 in the $+100$ to $+200$ range. Here realizing that $+10$ outcome is definitely unlucky (i.e., well below expectation). But it manages to avoid all those risky (negative yield) possibilities.
- *Question 3*: Is realizing the worst-possible result $|O^-|$ always unlucky?

  Answer: Not necessarily. For when all outcomes produce the same result—positive or negative—then none has a subexpectational yield.
- *Question 4*: Is optimal performance in a high risk situation always lucky?

  Answer: No—not when all outcomes are equally highly negative equivalent. But otherwise: Yes.

Chapter 8
Managing Luck

Luck divides the results of uncertain-outcome situations for a given individual into two main groups, the favorable and the unfavorable, the lucky (luck-positive) and the unlucky (luck negative). But regrettably, the proper answer to the question: What can I do to improve my luck *within* a *given* stochastic situation is: Nothing. Counting on luck is here problematic because there just is no way to affect outcomes where chance is the driver's seat. Once one is embarked in a specific chancy outcome situation the, die is cast. One simply cannot rely on the cooperation of luck. To change the bearing of luck one must change the situation. When chance is in the driver's seat the steering wheel is out of one's reach. But before this point is reached there are opportunities. In particular, improving one's skill is a decidedly effective way of diminishing risk. For when you improve your skill—and thus the win-probability in a particular mode of contest—this correspondingly diminishes your reliance on luck.

* * *

Seeing that luck pivots on incontrollable chance, "trusting to luck" is seldom a good idea. Skill, effort, contrivance, planning, and the like is almost always the more promising option. To be unthinking and heedless about getting into difficult situations and counting on luck to save the day is to invite trouble. Still, there are various constructive things one can do in relation to luck:

Inviting Good Luck The obvious precept "When possible, refrain from allowing yourself to get into chancy situations where luck is required to avert serious negativities" makes perfectly good sense. To maximize one's exposure to good luck: to give luck a chance by capitalizing on opportunities that put one into a situation where favorable developments can do one good. One cannot win the race one does not enter. Only by placing ourselves in a position where luck can do us some good can we hope to realize its benefits. By buying a ticket for the lottery can we possibly win it; by improving our qualifications we can increase the chances of securing a good job. Such measures increase the likelihood of a favorable outcome

N. Rescher, *Luck Theory*, Logic, Argumentation & Reasoning 20,
https://doi.org/10.1007/978-3-030-63780-4_8

and reduce our reliance on luck. The principles of prudential risk management are largely matters of simple common sense. And once our objectives are settled, prudence calls for keeping the odds on our side; luck is operative to the extent that success comes our way in the face of the odds. Either way—whether assessing prudence or adjudging luck, the determination of probabilities is a crucial factor.

Playing the Odds As Bishop Butler said, probability is [or should be!] "the guide of life." We can in many cases determine—or at least influence—the probability that good or bad luck will come our way. When you drive twice as far, you double the chance of an automobile accident. And while the drunk driver may not have an accident—the chances are that he may get away with it—still, if bad luck does come his way it does so more or less by invitation, since through his own action he has vastly increased the chances that something would go wrong.

It is almost always advisable to keep the odds on one's side by managing risks with reference to determinable probabilities—thereby also reducing the extent to which they place reliance on sheer luck. Wherever we can make a reasonable calculation of the expectations, we do well to be guided by this. Flying in the face of the odds is by its very nature imprudent. Once caught up in a chancy situation it is too late to influence luck: When chance is in the driver's seat and matters have slipped beyond one's reach. But before this point is reached there are opportunities.

Avoiding Undue Risks It is another cardinal rule of prudence to minimize one's exposure to bad luck: not to take undue or unnecessary chances. After all, people who do not court danger (who do not try to cross the busy roadway with closed eyes) need not count on luck to pull them through. The course of wisdom is to keep oneself out of harm's way. To avoid needless risks one's motto shall be "Do not push your luck": Do not do foolish, ill-advisedly risky things and count on good luck to rescue you from difficulties. The sagacious person avoids needless and excessive risks, keeping the odds in his favor so as to minimize the extent to which reliance need be placed on luck to save the day.

Buying Insurance It is another tried and true resource for impredictability management to provide some protection and buffering against the inevitable occurrences of unlucky eventuations; to build up some strategic reserves against misfortunes. People who take care to make proper provisions against unforeseeable difficulties by way of insurance, hedging, or the like. In effect, this involves forming a multiparticipatory syndicate where the many who do not lose out by the realization of an impredictable outcome cover the losses of those few who do, paying this price in advance to secure a like benefit for themselves should matters have so eventuated that they were the losers. By means of insurance, people are able to make a provision against impredictability by transforming the unpredictable outcome at issue from one of a large possible loss for a few into a small but assured sacrifice for many; namely, the cost of the insurance itself. While insurance does not alter the impredictability of outcome (one's house catching fire, one's ship foundering in a storm), it alters the situation so as to modify and offset the loss that would ensue if that unpredictable outcome were to eventuate unfavorably. (The process of "hedging one's bets"—for example by buying "futures," i.e., options for funds or for goods for future delivery—is closely analogous to this.) All such strategies (insurance, hedging and the like) are ways of coming to terms with impredictability and not ways of overcoming it. They provide the means for making the best of things in a predictively difficult world.

Extending Knowledge Clearly, the most promising way of handling impredictability due to ignorance rather than chance is to pursue further inquiries to eliminate or reduce the ignorance at issue to develop the information needed to arrive at prudently rational decisions instead of "trusting to luck." To be sure, the limited opportunities of the conditions of the place and time—and the costs and delays involved—mean that here we cannot always achieve as much as we would like. But doing what we can in this direction is generally worthwhile.

Improving One's Skill Skill enhancement is a decidedly effective way of diminishing risk. It may be true with respect to risk that "Luck favors the bold." However, if this is not because Reality somehow loads the dice and bends the choices in their favor, but only because rather than stay away from risk of any sort they are willing to seize chances where the eugaries are favorable. In particular, improving one's skill is a decidedly effective way of diminishing risk. For when you improve your skill—and thus the win-probability in a particular mode of contest this correspondingly diminishes your reliance on luck.

To be sure, skill-improvement never takes all luck out of it unless carried to the point of certain victory. On these issues, quantification of the matter once again serves to give an accurate picture of the underlying phenomenology.

Improving One's Chances Anything you can do to improve your win-probability all diminish the risk. Thus consider the by-now familiar situation of a binary win/lose situation:

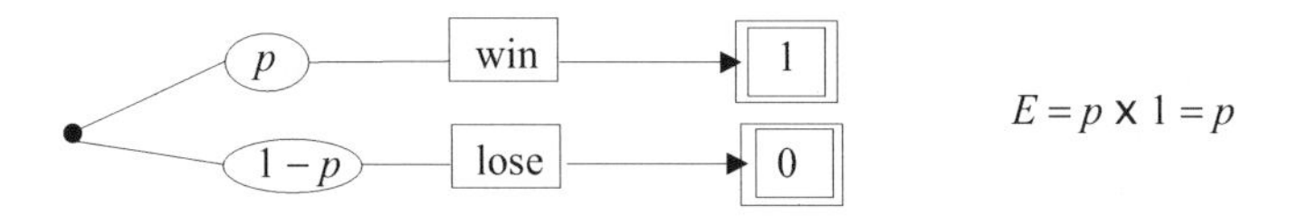

So here we have:

$$\lambda\{\text{win}\} = 1 - p$$

The bigger your win-probability the less your reliance on luck. And the relationship is linear. Increase the win-likelihood p by $Z\%$ in such situations and your luck-reliance decreases by exactly that percentage.

* * *

Skill improvement is paramount among the menus of luck management seeing that skill is a prime alternative to luck in the realization of favorable outcomes. And when someone succeeds in a chancy enterprise—be it a game or some more portentous venture—the question is often asked whether this happy result was due to skill or luck. This, of course, is something that can best be addressed on a statistical basis by comparing the proportion of cases in which the agent—or others at the same level of ability—has achieved a comparable success in the enterprise at issue.

Consider a game like Bridge or Dominoes where—unlike Chess or Go—success is in the main a *mixture* of skill and pure luck. Just how are the two to be divided?

Let it be that in such a game experts win over ordinarily competent players 80% of the time, while yet incurring 20% losses courtesy of "chance." And now consider four possibilities:

(A) an expert beats an expert
(B) an expert beats an ordinary player
(C) an ordinary player beats an ordinary player
(D) an ordinary player beats an expert

How much luck is needed in each case?

Note to begin with that those performance statistics indicate the four corresponding win probabilities to be: 50%, 80%, 50%, and 20%. Accordingly, the corresponding expectations of winning are .5, .8, .5, and .2. So with success scored as 1 our luck equation $\lambda = Y - E$ yields .5, .2, .5, and .8, respectively. In view of this, the ordinary players 20% success must be associated to luck. In the circumstances, the ordinary player who beats an expert is four times as lucky as the expert who turns the tables. So in 100 contests between them, we would expect the ordinary player to win 20% of the time. Should he in fact prevail only 10 times we would via $\lambda = Y - E$ deem him unlucky to the tune of 10 games. And should he win 25 times, we would ascribe 5 game excess to the luck that impels success beyond expectation.

* * *

Some theorists have luck and skill as jointly monopolizing the way to success, and envisions a linear continuum from pure skill to pure luck.[1] But this is no more than a plausible first approximation. For many other factors can enter into successful performance: persistence (in searching), health (in tennis), weight (in sumo wrestling), technology (in sail-boat racing), cheating (at cards), etc. The idea that a successful outcome is due either to skill or to luck—or some combination of the two—oversimplifies matters greatly. Other matters beyond skill can come into play.

The relation of luck and skill is complex. For example, one might think that performance tracks skill, so that when a contestant wins $X\%$ of contests he also wins $X\%$ of these wins by skill. But this cannot be. For suppose that in a contest between A and B, A's win record is represented by the fraction x. Then we will have the following situation with regard to proportionate shares.

	A's share	B's share
Total Wins	x	$1 - x$
Wins by skill	$x \times x$	$(1 - x) \times (1 - x)$
Wins by luck	$x - x^2$	$(1 - x) - (1 - x)^2$ $= x - x^2$

This would mean that, irrespective of the competence of the contestants, both have exactly the same number of wins by luck—i.e., both benefit from and depend upon luck to exactly the same extent. And this looks to be false in the very face of it.

Unraveling the ties between luck and skill can be difficult. In repetitive chancy situations like games or gambles, when outcome occurrences are distributed normally over the range of possibility then so is luck (seeing that $\lambda\{O\} = |O| - E$). On properly probabilistic grounds some performers are bound to do unusually well. So in any individual case there is no way of telling whether a given success was achieved by chance or by skill. It is a sad reality of life that in the statistical nature of things the attribution of success to skill rather than luck is always chancy.

* * *

Above and beyond achieving relevant expertise, there is yet another way of diminishing the role of luck in life—the rather radical remedy of the ancient Stoics and Epicureans, who recommended an indifference (*apathê*) that diminishes the range of things in which we take interest. This, of course, takes luck out of it since luck can play no role where no benefits or negativities are at stake. But for most of us this particular remedy comes at too high a cost.

Note

1. See Mauboussin 2012.

Appendix

Assessing Various Critiques

The quantitative theory of luck elaborated in this book was launched in my 1995 publication *Luck* (New York: Farrar-Straus-Giroux) and elaborated in the subsequent articles listed in the Bibliography. This has elicited a considerable response (also reported in the Bibliography) and evoked some proposals of alternative approaches. This appendix will address some of the critiques of the approach to luck measurement that the basis of this book's deliberations. As the following considerations will indicate, there is reason to think that some of these are based on oversights or misunderstandings, and some grounded in insufficient heed to the complexity of luck. For with luck as with many other assets (such as skill or riches) it is one thing to *be* and another to *feel*. And the deliberations to which the presently relevant investigations are addressed are not concerned with the *psychology* of luck but with what might be called its *economics*.

* * *

According to Steve D. Hales, "Rescher argues that only improbable events can be lucky or unlucky."[1] Not only is it wrong that Rescher so argues, but so doing would itself be patently wrong. The individual who emerges from a round of Russian Roulette realizes a result that is very probable, but yet can consider himself "darned lucky." And, who, as is only likely, finds himself one of the squad of twelve selected from among twenty for a suicide mission have every right to deem himself unlucky.

Steven D. Hales and Johnson want to draw a sharp contrast between "the *probability* theory [of luck]," according to which an occurrence is lucky (or unlucky) only if it was *improbable*, and the *modal* theory according to which an event is lucky only if it is *fragile*: "had the world been very slightly different—it would not have occurred."[2] There are three things amiss with this contention: (1) Someone can in fact be very lucky even if a probable event is realized. (In Russian roulette, as already indicted, or in "betting the farm" on the race's decided

N. Rescher, *Luck Theory*, Logic, Argumentation & Reasoning 20,
https://doi.org/10.1007/978-3-030-63780-4

favorite.) (2) There is no such thing as a slightly different world. Even the smallest change in the constitution of reality can and will ramify throughout (via the so-called "butterfly effect")[3] and eventually bring vast changes in its wake. (3)There is no plausible way of implementing the idea of "world-fragility" except by asking how likely it is that change *X* would occur if change *Y* had occurred, and thereby revert to probabilism.

To be sure, some theorists resist acknowledging luck in cases where the probable outcome is a godsend as with the Russian Roulette survivor, the "farm-betting" horserace gambler. They have it (with E. J. Coffman 2007, p. 392) that such a person is not actually lucky but "merely fortunate." However no-one has as yet explained how it is that these notions are so sharply delimited as to preclude these "fortunate" happenings from qualifying as lucky. Luck, after all, is not a clear-cut matter of yes or no, but one of extent. Certain occurrences transpire "with a little bit of luck," while others are enormously lucky, so that the matter is one of degree. (When you get as many dollars as the die shows points you are lucky to get three, but luckier still to get six.) Even when something probable happens, you can be lucky about it. When the hurricane damages 40% of the nearby houses but leaves yours untouched, you are one lucky householder.

* * *

Some discussants hold that two considerations suffice to make for luck: having an interest in an outcome and lacking control over its occurrence.[4] But counterexamples abound. Tomorrow's sunrise is nowise a matter of luck, and yet it something in which we have a strong interest but over which are have no control at all.

* * *

In his wide-ranging book *The Success Equation*, M. J. Mauboussin has is that: "Luck is a residual: it's what is left over after you've subtracted skill from outcome."[5] However, a good deal is wrong with this claim. (1) For one thing, luck and skill are not the only factors operative in producing a good (or bad) result: informativeness, persistence, resources, and a great deal else also comes into it. (If a commander's better plan had been inaugurated by a disgruntled subordinate, that battle is due neither to lack of skill nor to bad luck.) (2) For another, note that your skill at chess or tennis does not change from morning to afternoon and yet you may lose a game in the one and win in the other, not because your luck has changed but on the one occasion you faced a superior enemy. Mauboussin is right. Luck is a residual. But the does not issue from skill alone but potentially from any and all causal explanation whatever. Luck, in effect, is the default from occurrence-explainability at large.

* * *

In his informative discussion of luck, Andrew Latus has made the following objection:

> Rescher suggests that in order for . . . your being lucky to have some particular trait it must be possible for you to exist without that trait . . . He claims that in order for *you* to possess a particular trait you could have exited without that trait. But to image you without one of the

> traits that makes you your character is to imagine someone other than you. Clearly this aspect only . . . [holds] if each of the traits that make up a person's character is essential to that person.[6]

However, all I have maintained is that only *some* of a person's traits will qualify as essential/definitive/constitutive, while others can be contingent and thus potentially lucky. Thus some contingent circumstances (good health, for example) exemplify beneficial conditions not realized by luck, whereas wealth due to a found treasure would be decidedly lucky. Accordingly, it fails to follow "that Rescher has misunderstood the action of luck . . . [because] constitutive luck is not an incoherent notion [as he erroneously believes]."[7] On my principles there is no need to deny the prospect of "constitutive" luck, but only maintain that the constitutive factors properly deemed lucky cannot be among those viewed as essential for the individual's being as is.

* * *

In an interesting discussion of the psychology of luck J. L. Dessalles offers two objections to the idea that the amount of luck in binary, win-lose situations can be measured by $\lambda = \Delta(1 - p)$.[8]

The first is that "it does not distinguish moderately unlucky outcomes from highly unlucky ones, as both would provide emotion roughly equal to Δ," this Δ being "the difference that the occurrence makes for the interests at stake." However, this critical complaint has three flaws: (1) the formula's factor of $(1 - p)$ is exactly designed to accommodate the difference between likely and unlikely outcomes, and (2) the idea that both lucky and unlucky outcomes would normally evoke the same "emotional response" is questionable and problematic. (Are there no such things as "lucky escapes" or "long shots"?) (3) The metric theory of luck does not revolve about the matter of "providing emotion," which is substantially beside the point of present purpose.

Dessalles' second objection is that the luck formula "fails to capture the crucial presence of a counterfactual [considerations] that the amplitude [presumably he means *quantity*] of luck is controlled by the "distance" to an alternative [unlucky] outcome." But this objection is dissolved by the fact that the contrasting outcome is built into the very constitution of that Δ in the luck formula. Thus consider the paradigm illustration of Display A. It is correct to say that the luck of winning \$1 in Case I "is controlled by the 'distance' to an alternative outcome." But just this fact is captured by the consideration that via $\lambda = Y - E$ the amount of win-luck in Case I is $3(1 - p)$, and in Case 2 is the substantially greater $11(1 - p)$, exactly for the reason specified in that complaint. On this basis, the answer to the question "How lucky is it to win a dollar with probability p—how much luck is involved here?" differs in the two cases. And it arises in a way that illustrates rather than counters the counterfactual that

Display A

A COMPARISON CASE

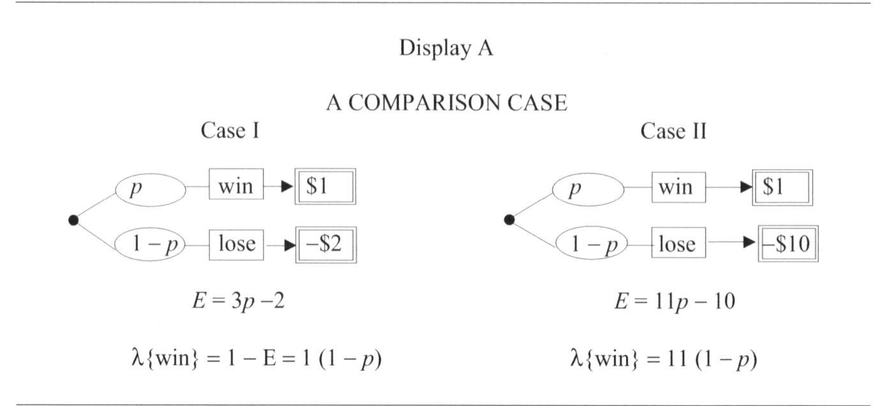

"If the loss-yield had been different, the luck of the $1 win would have been different." That Basic Luck Formula provides for the "crucial presence of a counterfactual" rather than conflicting with.

* * *

Patrick Beach has poses the following objection:

> Rescher's formula [for binary win/lose luck: $\lambda = \Delta(1 - p)$] does not recognize that the probability of an event is sensitive to what background information we hold as relevant. Thus there are different vantage points for which a given event for a given agent is lucky and other vantage points for which it is unlucky. Rescher's formula has no mechanism for relativizing the luck of an event to different vantage points and so is insensitive to the various values that can reasonably be assigned to the probability of an event.[9]

Yet consider an analogous issue. Ronald's car was in a collision with a black sports car. Was he lucky because 80% of black cars are insured or unlucky because only 20% of sports car are? No formula in the world can—or can be expected to—tell us which probability is to be used here, whether that based on the 80% statistic or that based on the 20% one.[10] But in this regard, the luck equation is exactly like any other parametric formula. Take the area of a rectangle, subject to the equation: Area = Base × Height. Of course such a formula does not purport to tell how those qualities are to be measured. Its bearing is clearly conditional: IF the base is such-and-such AND the height is so-and-so, then the area will be the product of the two. Determining those parametric values themselves presents an entirely different issue. The fact that the formula doesn't resolve it is no basis for to complaint. And exactly the same as in luck formula

$$\lambda\{O\} = p\{O\} - E \text{ or in the binary case} : \lambda = \Delta(1 - p)$$

That the formula does not measure those parametric values but requires their input to be provided by other means is only natural and to be expected.

And the same situation obtains with the other instrumentation of rational decision theory. Expected value calculations on risk minimization rules all call for probabilities and utilities to be produced *ab extra* and antecedently before their parametric formulas can do their interrelational work. And as far as Luck Theory is concerned the numerical values of its parameters (probabilities, yields, expectations) are substantially objective factors pertaining to issues of rational decision theory at large.

* * *

In his instructive essay on "Thinking about Luck," E. J. Coffman argues that luck must be improbable, that "if the odds actually favored a given event's arrival . . . then its occurrence was hardly a stroke of *luck* for you."[11] Against this view I could cite not only the aforementioned case of the person who survived 1-chamber Russian Roulette, but also, the individual left untouched by a plague that killed 30% of his neighbors, or the soldier whose unit suffered 40% battle causalities. To operate with a concept of luck that precludes probable outcomes from qualifying as lucky is to have a decidedly eccentric conception of the matter.

* * *

Another complaint that has been lodged against the measurement of luck in terms of probability and yield is that "Not all good luck evaluations imply a benefit or gain."[12] But this fails to distinguish matters adequately. For while there can indeed be luck without certain gain (as when the status quo is maintained in the face of threatening disaster) this sort of thing is always a matter of benefit. So the objection overlooks the fact that the avoidance of negativities will also qualify as positive. After all, narrow escapes are paradigmatic examples of beneficial good luck that may come at a price.

Moreover, this critic's objection that "Rescher's claim that degree of luck is simply a function of outcome attractiveness, on the one hand, and low value probability, on the other"[13] overlooks the fact (a) that I speak of an outcome's *yield* rather than "its "attractiveness," and (b) that even were it otherwise, the fact remains that "attractiveness" can be both positive and negative, so that unattractiveness must also be reckoned with.

K. H. Teigen sees it as a problem for the present approach to luck measurement that "Our studies indicate that [for luck] a high probability of a counterfactual outcome is even more crucial than a low probability for what actually took place."[14] But consider the contrast between an actual occurrence O and its contrary-to-fact alternative, not-O. If the probability of the one is p, then that of the other will of course automatically be $1 - p$. So if one is higher, the other is automatically and inevitably lower. It is hard to see how one or two inevitably coordinated factors can possibly manage to be "even more crucial" than the other.

* * *

In his interesting discussion of luck, K. H. Teigen maintains that the win/lose luck formula $\lambda\{O\} = \Delta \times (1 - p\{O\})$ (when Δ = [non - yield − lose - yield]) fails in certain cases to capture people's luck assessment because "Not all [winning] events with the same probability are equally lucky."[15] But of course the formula does not affirm otherwise, but takes the factor of yield as being of the essence.

As an improved measurement of luck Teigen proposes a counterfactually geared replacement based on the idea that a suitable measure[16] is given by

$$\mathcal{L} = c(U(f) - U(Cf))$$

where c is a situationally determined constant, $U(f)$ is the "utility" or value of the outcome-occurrence f, and $U(Cf)$is to be "the utility of the counterfactual outcome Cf "that would occur if (contrary to fact) f were not the case" (p. 143). But this version counterfactual approach is in the end nothing new or different from ours. For in our notation that binary $\mathcal{L}$-formula would be written as

$$\mathcal{L}\{f\} = c(|f| - |\text{not-}f|)$$

Where c is again an appropriate outcome-correlative constant. But this is simply a special case falling under our principle $\lambda = Y - E$. For consider

$$\begin{aligned} \lambda\{f\} &= |f| - [p \times |f| + (1 - p) \times |\text{not-}f|)] \\ &= (1 - p)(|f| - |\text{not-}f|)] \end{aligned}$$

Thus $\mathcal{L}$ here corresponds exactly to λ. So on present principles that proposed alternative is nothing new, and moreover instead of an otherwise mysterious c we now can have the clearly meaningful outcome improbability $1 - p$.

* * *

Some theorists maintain that, as one writer puts it, when such unforeseeable events as unexpected benefits or losses occur, "our agent may well be inclined to put this event down to luck, and it is not this judgement that should concern us (as it is almost certainty made when not being in possession of full information), but rather what we would say about the situation once we are apprised of all the relevant facts."[17] However this is very problematic both because luck is ultimately a matter of how things stand for the protagonist's—not opinions but—interests, something for which he is frequently but not always the best judge. Moreover, and more importantly, it once more deserves note that in this world we are seldom, if ever, "appraised of all the relevant facts."

* * *

An agent can be considered lucky when benefitting from a by-him unforeseeable occurrence even if this were securely predictable by others. (A protagonist not only *feel* but *is* lucky in winning at Roulette when (unbeknownst to the player) the wheel is fixed: or in coming upon a treasure which, unbeknownst to him, has been deliberately placed in the way.) Accordingly, one should acknowledge epistemic as well as stochastic chance, and envision a correspondingly duality of subjective and objective luck. The protagonist of "The Lady and the Tiger" must be deemed lucky if successful. Still, various critics have called this into question,[18] insisting that the inherent chanciness of luck is at odds with the benign results of "deliberate contrivance by others" sort that while "such events may seem lucky, they in fact are not."[19] But this line rides roughshod over the distinction between subjectively psychological luck on the one side and objectively ontological luck on the other with the former regulated to the trash heap of what is seeming and apparent. This flies in the face of the circumstance that attributions of luck standardly approach the matter from the angle of the concerned agent's standpoint and not that of omniscience. Perhaps if nature were a theater of Laplacean determinism, chance might be an illusion and there would be no luck "from the standpoint of the universe." But even then, epistemic luck could survive intact and fully deserve its name.

* * *

Michael J. Zimmerman has it that

> Luck is something that occurs beyond one's control. I am not using it as some do (for example, Nicholas Rescher) as follows: something that occurs as a matter of luck is something that occurs by chance, that is, something real where there is or was some possibility of its not occurring.[20]

Of course people are free to use words as they wish—but this can easily exact a price in understanding when it falls outside the usual and familiar range. And in the present case this approach to the matter seems very questionable. Uncontrollability is neither necessary nor sufficient for luck. It is not sufficient because we cannot control the rising of the sun and the moon or the standard succession of weekdays, but they are certainly not matters of luck. And it is not necessary because we do control the output of our typewriter but that does not prevent misprints from qualifying as bad luck. The helmsman was still in control of the ship when the Titanic hit the iceberg. The driver who accidently turns the wrong way onto a one-way street fully controls the vehicles movement but may yet be unlucky about it

Control precludes luck only when it is *knowingly, deliberately and appropriately and effectively* exercised. The person who operates the controls of device or process without fully understanding it may well have things go as desired but if so will be lucky indeed. To reemphasize, the crux of luck is not the absence of control, but the absence of reliable foreseeability.

* * *

Some theorists maintain luck and skill monopolizes the way to success, and envisions a linear continuum from pure skill to pure luck.[21] This is certainty a plausible first approximation. In the end, however, many other factors can enter into successful performance: persistence (in searching), health (in tennis), weight (in sumo wrestling), technology (in sail-boat racing), cheating (at cards), etc. The idea that a successful outcome is due either to skill or to luck oversimplifies matters greatly.

* * *

In his informative study of constitutive luck, Andrew Latus has it that:

> A problem arises with the chance account [of luck] . . . Suppose I have the ability to bring about a rare event of great value and I make use of this ability. An unlikely, valuable event

> has therefore taken place; but it does not seem correct to describe me as lucky that the event has occurred.[22]

This objection, however, does not touch the heart of the matter. After all, the postulated result did not happen by luck but deliberate effort on the basis of planning and creative competence, with unforeseeability out of the picture. And on present principles the successful realization of an expectable outcome has zero luck, notwithstanding whatever benefit it may provide.

* * *

In his informative article on "A Measure of Luck" specifically oriented at the game of Backgammon, Doyle Zare proposed the "alternative" that here "luck is [determined by] the equity you gain through the roll of the dice": it is "the equity of your best possible play minus your equity before the roll of the dice."[23] That crucial measure of "equity" is not addressed in Zare's paper, but clearly the only "equity" one can have—before the dice are rolled—is the mathematical expectation. And "best possible play" is presumably that associated with the actually best possible outcome (rather than what is indicated as such by the *available* information.). On this construal we would have it that:

$$L = |O^{+}| - E$$

Thus luck addressed in Zare's discussion is in effect, the luck of winning—of achieving the best possible outcome O^{+}. Our present approach via $\lambda = Y - E$ would, of course, lead to:

$$\lambda\{O^{+}\} = |O^{+}| - E$$

So given the construal-proposals postulated above with regard to Zare's approach, his luck measure actually yields exactly the same result envisioned by the present approach for optimal outcomes.

* * *

Overall, it would appear that all of the critiques and modifications to the present approach to luck-assessment presented in the recent literature can either be accommodated or rebutted in this setting.

* * *

In the end, the present treatment of luck has to address the following problem that can be posed in the following terms:

> Luck as people usually deal with it is an informal, unsystematic, and generally unquantified idea. What then enables a formal and quantitative treatment such as the presently proposed to qualify as a theory of *luck*?

The answer roots in an analogy. Arithmetic as originally managed dealt with was a matter of purely mental calculation with small numbers. As subsequently developed via formal procedures and elaborate (complex) routines, we still claim to be dealing with arithmetic in that original sense. And the justification for this lies in the fact that those formalized proceedings harmonize and agree with the informal presystematic results. The results are concordant and uniform throughout the range of overlap between the systemation and the pre-systemic throughout this common region intuition and formalization agree in harmonious unison. And just this aspect of the formalization of arithmetic is also at hand in the present formalization of luck theory.

Notes

1. Steven D. Hales, 2016, p. 490.
2. Hales/Johnson in Pritchard & Whittington 2015, p. 50.
3. On the butterfly effect see the Wikipedia article of this title where ample references for further literature is provided.
4. See, for example, Broncour-Berroucel 2015, p. 22.
5. Mauboussin 2012, p. 16.
6. Latus 1998, p. 42.
7. Op. cit.
8. See Dessalles 2010, p. 1.
9. Beach 2012, p. 59.
10. Actually it is neither: we need conjoint illustrations for an appropriate estimation of probability.
11. Coffman 2007 and compare Prichard 2005.
12. K. H. Teigen in Mandel et al. 2005, p. 132.
13. Op. cit, p.135.
14. Teigen 2019, p, 349.
15. Teigen in Mandel et. al. 2005, p. 131.
16. Teigen in D. R. Mandel et. al. 2005, p. 143.
17. Prichard 2005, p. 119.
18. E.g., it is "far from clear that this [sort of thing] is a case of luck, however, no matter how well the agent may regard it as such." (Prichard 2005, p. 215.)
19. See Lackey 2008, p. 263.
20. Zimmerman 2002, p., 559n.
21. See Mauboussin 2012.
22. Latus 2003, p. 467.
23. Zare 2000, p. 2.

Bibliography

Aldous, David, "On Chance and Unpredictability: Lectures on the Links between Mathematical Probability and the Real World" (lecture notes: https://case.edu/artsci/math/esmeckes/ussy292d/aldous_2015.pdf, 2015).

Andre, Judith, "Nagel, Williams, and Moral Luck," *Moral Luck*, ed. Daniel Statman (State University of New York Press, 1993), pp. 123–130.

André, Nathalie, "Good Fortune, Luck, Opportunity and Their Lack: How do Agents Perceive Them?" *Personality and Individual Differences*, vol. 40 (2006), pp. 1461–1472.

Ashley, R. N., *The Wonderful World of Superstition, Prophecy, and Luck* (New York: Dembner Books, 1984).

Ballantyne, Nathan, "Anti-luck Epistemology, Pragmatic Encroachment, and True Belief," *Canadian Journal of Philosophy*, vol. 41 (2011), pp. 485–504.

Ballentyne, Nathan, "Does Luck have a Place in Epidemology?" *Synthese*, vol. 191 (2014). pp. 1391–1407.

Ballentyne, Nathan, "Luck and Interest," *Synthese*, No. 185 (2012), pp. 219–334.

Bandura, Albert, "The Psychology of Chance Encounters and Life Paths," *American Psychologist*, vol. 37 (1982), pp. 747–755.

Beach, Patrick A., *Moral Luck* (Syracuse: University of Syracuse, Doctoral Dissertation, 2012).

Blancha, David, *The Moral of Luck* (New York: Columbia University, Doctoral dissertation, 2015).

Bohm, David, *Causality and Chance in Modern Physics* (Princeton, Princeton University Press, 1957).

Broncano-Berrocal, Fernando, "Luck as Risk and the Lack of Control Account of Luck," *Metaphilosophy*, vol. 46 (2015), pp. 1–25, also in Pritchard and Whittington, (2015), pp. 3–26.

Browne, Brynmor, "A Solution to the Problem of Moral Luck," *Philosophical Quarterly*, vol. 42 (1992), pp. 345–356.

Chinnery, Ann, "Constitutive Luck and Moral Education," in M. Moses (ed.), *Philosophy of Education* (Urbana-Champaign, IL: Philosophy of Education Society), pp. 45–53.

Church, Ian M. and Robert J. Hartman, *The Routledge Handbook of the Philosophy and Psychology of Luck* (New York & London: Routledge, 2019).

Coffman, E. J., "Thinking about Luck," *Synthese*, vol. 158 (2007), pp. 385–398.

Coffman, E. J., *Luck: Its Nature and Significance for Human Knowledge and Agency* (New York: Palgrave Macmillan, 2015).

Cohen, John, *Chance, Skill, and Luck: The Psychology of Guessing and Gambling* (Harmondsworth UK: Penguin Books, 1960).

N. Rescher, *Luck Theory*, Logic, Argumentation & Reasoning 20,
https://doi.org/10.1007/978-3-030-63780-4

Darke, Peter R., and Jonathan L. Freedman, "Lucky Events and Beliefs in Luck: Paradoxical Effects on Confidence and Risk-Taking," *Personality and Social Psychology Bulletin*, vol. 23 (1997a), pp. 378–388.
Darke, Peter R., and Jonathan L. Freedman, "The Belief in Good Luck Scale," *Journal of Research in Personality*, vol. 31 (1997b), pp 486–511.
DeGrafe, J. A. M., *Luck and Justification*, (Gronigen: University of Gronigen, Doctoral Dissertation, 2017).
Dessalles, J. L. "Emotions in Good Luck and Bad Luck," *Proceedings of the 32nd Annual Conference of the Cognitive Science Society* (Portland, PR, 2010). [Available on the internet.]
Edgeworth, F. Y., *Mathematical Psychics*, (London, 1881; reprinted New York, 1961).
Engle Jr., Mylan, "Is Epistemic Luck Compatible with Knowledge," *The Southern Journal of Philosophy of Philosophy*, vol. 30 (1992), pp. 59–75.
Enoch, David and Andrei Marmor, "The Case against Moral Luck." *Law and Philosophy*, vol. 26 (2007), pp. 405–436.
Enoch, David and Ehud Guttel, "Cognitive Biases and Moral Luck," *Journal of Moral Philosophy*, vol. 7 (2010), pp. 372–386.
Foley, Richard, "Epistemic Luck and the Purely Epistemic," *American Philosophical Quarterly*, vol. 21 (1984), pp. 113–124.
Fringer, Rob A. and Jeff K. Lane, *Theology of Luck: Fate, Chaos, and Faith* (Kansas City, MO: Beacon Hill Press, 2015).
Giberson, Karl W., *Abraham's Dice: Chance and Providence in the Monotheistic Traditions* (New York: Oxford University Press, 2016).
Gladwell, Malcolm, *Outliers: Why Some People Succeed and Some Don't* (New York: Little Brown & Co., 2008).
Greco, John, "A Second Paradox Concerning Responsibility and Luck," *Metaphilosophy*, vol. 26 (1995), pp. 81–96.
Gunther, Max, *The Luck Factor* (Boston: Harriman House, 1977).
Hacking, Ian, *The Emergence of Probability* (Cambridge: Cambridge University Press, 1975).
Haji, Ishtiyaque, "Libertarianism, Luck, and Action Explanation," *Journal of Philosophical Research*, vol. 30 (2005), pp. 321–340.
Hales, Steven D., "Why Every Theory of Luck is Wrong," *Noûs*, vol. 50 (2016), pp. 490–508
Hales, Steven D. and Jennifer A. Johnson, "Cognitive Bias and Disposition in Luck Attribution," in Pritchard and Whittington 2015.
Hall, Barbara J., "On Epistemic Luck," *The Southern Journal of Philosophy*, vol. 32 (1994), pp. 79–84.
Harper, William, "Knowledge and Luck," *The Southern Journal of Philosophy*, vol. 34 (1996), pp. 273–283.
Hartmann, Robert (ed.), *Routledge Handbook of the Philosophy and Psychology of Luck* (London: Routledge, 2019).
Hawthorne, John, *Knowledge and Lotteries* (Oxford: Oxford University Press, 2004).
Hayano, David M., "Strategies for the Management of Luck and Action in an Urban Poker Parlour," *Urban Life*, vol. 6 (1978), pp. 475–488.
Hay-Gibson, Naomi, "A River of Risk: A Diagram of the History and Historiography of Risk Management," Interdisciplinary Studies in the Built and Virtual Environment, vol. 1 (2008), 149–158.
Honoré, A. M., "Responsibility and Luck," *The Law Quarterly Review*, vol. 104 (1988), pp. 530–553.
Hurley, Susan, *Justice, Luck, and Knowledge* (Cambridge: Harvard University Press, 2003).
Jeffrey, Richard C., *The Logic of Decision* (New York: McGraw-Hill, 1983).
Jensen, Henning, "Morality and Luck," *Philosophy*, vol. 59 (1984), pp. 323–30.
Kahnemann, Daniel, Paul Slovic, and Amos Tversky (eds.) *Judgment Under Incertainty: Heuristic and Biases* (Cambridge: Cambridge University Press, 1982).

Kahnemann, Daniel and Carol A. Varey," "Propensities and Counterfactuals: The Loser that Almost Won," *Journal of Personality and Social Psychology*, vol. 59 (1990), pp. 1101–1110.
Kessler, Kimberly, "The Role of Luck in the Criminal Law," *University of Pennsylvania Law Review*, vol. 142 (1994), 2183–2237.
Keyes, Ralph, *Chancing It: Why we Take Risks* (Boston and Toronto: Little Brown, 1985)
King, Mervyn A. and J. A. Kay, *Radical Uncertainty: Decision-Making for an Unknowable Future* (New York: Norton, 2020).
Knight, Frank, *Risk, Uncertainty, and Profit* (Boston, Houghton-Mifflin, 1921).
Knight, Frank, *Luck Egalitarianism: Equality, Responsibility, and Justice* (Edinburgh: Edinburgh University Press, 2009).
Lackey, Jennifer, "What Luck Is Not," *Australasian Journal of Philosophy* vol. 86 (2008), pp. 255–267.
Latus, Andrew M., *Avoiding Luck: The Problem of Moral Luck and its Significance* (Toronto: University of Toronto Doctoral Dissertation, 1998).
Latus, Andrew M., "Moral and Epistemic Luck," *Journal of Philosophical Research*, vol. 25 (2000), pp. 149–172.
Latus, Andrew, "Constitutive Luck," *Metaphilosophy*, vol. 34 (2003), pp. 460–475.
Lennox, James, "Aristotle on Chance," *Archiv für Geschichte der Philosophie*, vol. 66 (1984), pp. 52–60.
Levy, Neil, *Hard Luck: How Luck Undermines Free Will and Responsibility* (New York: Oxford University Press, 2011).
Luce, R. D., and Howard Raiffa, *Games and Decisions: Introduction and Critical Survey* (New York: Wiley, 1957).
Makridakis, Spyros G., Robin M. Hogarth, and Anil Gaba, *Dance with Chance: Harnessing the Power of Luck* (Oxford: Oneworld, 2009).
Mandel, David R., Denis J. Hilton and Patrizia Catellani, *The Psychology of Counterfactual Thinking* (London & New York: Routledge, 2005).
Mauboussin, Michael, *The Success Equation: Untangling Skill and Luck in Business, Sports, and Investing* (Boston: Harvard Business Review Press, 2012).
Mele, Alfred, *Free Will and Luck* (New York: Oxford University Press, 2006).
Meyer, John P., "Causal Attributions for Success and Failure," *Journal of Personality and Social Psychology*, vol. 38 (1980), pp. 704–715.
Mlodinow, Leonard, *The Drunkard's Walk: How Randomness Rules Our Lives* (Harmondsworth UK: Penguin Books, 2008).
Morillo, Carolyn, "Epistemic Luck, Naturalistic Epistemology, and the Ecology of Knowledge," *Philosophical Studies*, vol. 46 (1984), pp. 109–129.
Nagel, Thomas. "Moral Luck," *Moral Luck*, ed. Daniel Statman (State University of New York Press, 1993), pp. 57–72.
Nussbaum, Martha, *The Fragility of Goodness: Luck and Ethics in Greek Tragedy and Philosophy* (Cambridge: Cambridge University Press, 1986).
Nussbaum, Martha C., "Luck and Ethics," *Moral Luck*, ed. Daniel Statman (Albany, State University of New York Press, 1993), pp. 73–108.
Oates, Wayne E., *Luck: A Secular Faith* (Louisville, KY: Westminster John Knox Press, 1995).
Pritchard, Duncan, "Epistemic Luck," *Journal of Philosophical Research*, vol. 29 (204), pp. 193–222.
Pritchard, Duncan, *Epistemic Luck* (Oxford: Clarendon Press, 2005).
Pritchard, Duncan, "The Modal Account of Luck," *The Philosophy of Luck*, ed. Duncan Pritchard and Lee John Wittington (Wiley Blackwell 2015), pp. 143–167.
Pritchard, Duncan, "Sensitivity, Safety, and Anti-Luck Epistemology," *The Oxford Handbook of Skepticism* (Oxford University Press, 2008), pp. 438–455.
Pritchard, Duncan and Lee John Whittington (eds.), *The Philosophy of Luck* (Oxford: Wiley-Blackwell, 2015).

Pritchard, Duncan and Michael Smith, "The Psychology and Philosophy of Luck," *New Ideas in Psychology*, vol. 22 (2004), pp. 1–28.
Rapport, Anatol, *Fights, Games, and Debates* (Ann Arbor, University of Michigan Press, 1960).
Raz, Joseph, "Agency and Luck," *Columbia Public Law & Legal Theory Working Papers*," Paper 9170 (2009).
Rescher, Nicholas, *Introduction to Value Theory* (Englewood Cliffs: Prentice Hall, 1969).
Rescher, Nicholas, "Luck," *Proceedings of the American Philosophical Association*, vol. 64 (1990), pp. 5–19.
Rescher, Nicholas, "Moral Luck," *Moral Luck*, ed. Daniel Statman (State University of New York Press, 1993), pp. 141–166.
Rescher, Nicholas. *Luck: The Brilliant Randomness of Everyday Life* (New York: Farrar-Strass-Giroux, 1995; reprinted in 2000 by the University of Pittsburgh Press.)
Rescher, Nicholas, "The Machinations of Luck," *Metaphilosophy*, vol. 45 (2014), pp. 620–626.
Rescher, Nicholas, "The Probability Account of Luck," by Church and Hartman (2019).136–147.
Richards, Norvin, "Luck and Desert," *Mind*, vol. 95 (1986), pp. 198–209.
Riggs, Wayne, "Luck, Knowledge, and Control," *Epistemic Value,* eds. A. Haddock, A. Millar, and D. H. Prichard (Oxford University Press, 2009a), pp. 204–221.
Riggs, Wayne, "Luck, Knowledge, and 'Mere' Control," *Epistemic Value,* eds. A. Haddock, A. Millar, and D. H. Prichard (Oxford University Press, 2009b), pp. 177–189.
Rosebury, Brian, "Moral Responsibility and 'Moral Luck'," *The Philosophical Review*, vol. 104 (1995), pp. 499–524.
Schellinger, Daniel, *The Politics of Luck* (Toronto: University of Toronto, Doctoral dissertation, 2018).
Schinkel, Anders, "The Problem of Moral Luck: An Argument against its Epistemic Reduction," *Ethical Theory and Moral Practice*, vol. 12 (2009a), pp. 267–277.
Segall, Shlomi, *Health, Luck, and Justice* (Princeton, NJ: Princeton University Press, 2010).
Schinkel, Anders, "The Problem of Moral Luck: An Argument against its Epistemic Reduction," *Ethical Theory and Moral Practice*, vol. 12 (2009b), pp. 267–277.
Sorenson, Roy, "Logical Luck," *Philosophical Quarterly*, vol. 48 (1998), pp. 319–334.
Sproul, R. C., *Not a Chance: The Myth of Chance in Modern Science and Cosmology* (Grand Rapids, MI: Baker Books, 1994).
Statman, Daniel, "Moral and Epistemic Luck," *Ratio*, vol. 4 (1991), pp. 146–56.
Statman, Daniel, *Moral Luck* (Albany: SUNY Press, 1993).
Steglich-Petersen, Asbjørn, "Luck as an Epistemic Notion," *Synthese*, vol. 176 (2010), pp. 361–377.
Sverdlik, Steven, "Crime and Moral Luck," *American Philosophical Quarterly*, vol. 25 (1988), pp. 79–95.
Taleb, Nassim N., *Fooled by Randomness: The Hidden Role of Chance in Life* (New York: Random House, 2004).
Taleb, Nassim N., *The Black Swan: The Impact of the Highly Improbable* (New York: Random House, 2007).
Teigen, Karl H., "Luck: The Art of a Near Miss," *Scandinavian Journal of Psychology*, vol. 37 (1996), pp. 156–171.
Teigen, K. H., "Luck and Risk," in Church and Hartman (2019), pp. 345–355.
Thompson, Edmund R and Gerard P. Prendergast, "Belief in Luck and Luckiness: Conceptual Clarification and New Measure Validation," *Personality and Individual Differences*, vol. 37 (2013), pp. 501–506.
Thomson, Judith Jarvis, "Morality and Bad Luck," *Metaphilosophy*, vol. 20 (1989), pp. 203–221.
Thomson, Judith Jarvis, "Morality and Bad Luck," *Moral Luck*, ed. Daniel Statman (State University of New York Press, 1993), pp. 195–216.
Todhunter, Isaac, *A History of the Mathematical Theory of Probability from the Time of Pascal to that of Laplace* (Cambridge: Cambridge University Press, 2014).

von Neumann, John, and Oskar Morgenstern, *Games and Decisions* (New York: J. Wiley & Sons, 1944).
von Neumann, John, and Oskar Morgenstern, *Theory of Games and Economic Behavior* (Princeton: Princeton University Press, 1946; 2nd ed. 1953).
Vahid, Hamid, "Knowledge and Varieties of Epistemic Luck," *Dialectica*, vol. (2001), pp. 351–362.
Wagenaar, W. A., *Paradoxes of Gambling Behaviour* (Hillsdale: Erlbaum, 1988).
Wagenaar, W. A., and G. B. Keren, "Chance and Luck are not the Same," *Journal of Behavioral Decision Making*, vol. 1 (1988), pp. 65–75.
Walker, Margaret Urban, "Moral Luck and the Virtues of Impure Agency," *Moral Luck*, ed. Daniel Statman (State University of New York Press, 1993), pp. 235–250.
Weaver, Warren, *Lady Luck: the Theory of Probability* (Garden City, New York: Anchor, 1963).
Williams, Bernard, *Moral Luck* (Cambridge: Cambridge University Press, 1981).
Williams, Bernard, "Postscript," *Moral Luck*, ed. Daniel Statman (State University of New York Press, 1993), pp. 251–258.
Williams, Bernard and T. Nagel "Moral Luck," *Proceedings of the Aristotelian Society*, Supplementary Volumes, vol. 50 (1976), pp. 115–151.
Wiseman, Richard, *The Luck Factor* (London: Random House, 2003).
Wiseman, Richard and C. Watt, "Measuring Superstitious Belief: Why Lucky Charms Matter," *Personality and Individual Differences*, vol. 37 (2004), pp. 1533–1541.
Wolf, Susan, "The Moral of Moral Luck," *Philosophical Exchange*, vol. 31 (2001), pp. 5–19.
Zagzebski, Linda, "Religious Luck," *Faith and Philosophy*, vol. 11 (1994), pp. 397–413.
Zare, Douglas, "A Measure of Luck [in Backgammon]," *GammonVillage* (https://www.gammonvillage.com/index.cfm), 2000.
Zimmerman, Michael J., "Luck and Moral Responsibility," *Ethics*, vol. 97 (1987), pp. 374–86
Zimmerman, Michael J., "Taking Luck Seriously," *Journal of Philosophy,* vol. 99 (2002), pp. 553–576.

Name Index

N. Rescher, *Luck Theory*, Logic, Argumentation & Reasoning 20,
https://doi.org/10.1007/978-3-030-63780-4

Printed in the United States
by Baker & Taylor Publisher Services